Deterministic Flexibility Analysis

Analysis

Theory, Design, and Applications

Deterministic Flexibility Analysis

Theory, Design, and Applications

By

Chuei-Tin Chang

Vincentius Surya Kurnia Adi

CRC Press
Taylor & Francis Group
Boca Raton London New York

CRC Press is an imprint of the
Taylor & Francis Group, an **informa** business

CRC Press
Taylor & Francis Group
6000 Broken Sound Parkway NW, Suite 300
Boca Raton, FL 33487-2742

First issued in paperback 2019

CRC Press is an imprint of Taylor & Francis Group, an Informa business

No claim to original U.S. Government works

ISBN-13: 978-1-4987-4816-2 (hbk)
ISBN-13: 978-0-367-87541-1 (pbk)

Visit the Taylor & Francis Web site at
http://www.taylorandfrancis.com

and the CRC Press Web site at
http://www.crcpress.com

Contents

Preface

This book presents an introduction to both theoretical and application aspects of flexibility analysis, which is primarily concerned with the task of assigning a sound measure of the operational capability of a given system under the influence of uncertainties. The formal definitions of several available performance indices, their mathematical formulations, and the corresponding algorithms and codes are provided in sufficient detail to facilitate implementation. It is therefore appropriate for an industrial reference, a senior-level design course, or a graduate course in chemical process analysis.

Traditionally, design and control decisions are made in sequential stages over the life cycle of a chemical plant. In the design phase, the "optimal" operating conditions and the corresponding material and energy balance data are established mainly on the basis of economic considerations. In the subsequent step, the control systems are configured to maintain the key process conditions at the fixed nominal values. Because it is often desirable to address the operability issues at the earliest possible stage before stipulation of control schemes, the systematic incorporation of flexibility analysis in process synthesis and design has received considerable attention in recent years. This book focuses to a large extent on computation methods for evaluating *deterministic* performance measures, that is, the steady-state, volumetric, dynamic, and temporal flexibility indices in various applications. The contents in each chapter can be briefly summarized as follows:

Chapter 1 provides a general introduction to flexibility analyses and qualitative definitions of several different measures of system resiliency.

Chapter 2 presents the conventional *steady-state* flexibility index (FI_s), which is useful for characterizing the continuous processes. The corresponding model formulation, the existing solution strategies, and a simple example are included therein to ensure thorough understanding. The undesirable possibilities of misrepresentation, which may be attributed to the off-center nominal points and/or the nonconvex feasible regions, are discussed at the end of the chapter.

These drawbacks in the steady-state flexibility analysis can be circumvented by using a different metric, that is, the volumetric flexibility index (FI_v) discussed in Chapter 3. The geometric interpretation of FI_v and the required algorithms are given in this chapter to help readers understand conceptually and carry out specific computations. Several case studies are also presented to demonstrate the benefits of this alternative approach.

Because both steady-state and volumetric flexibility indices are suitable only for gauging the continuous processes operated at steady states, the batch systems should be evaluated differently. By replacing the equality constraints of the original steady-state model with a system of differential algebraic equations (DAEs), a modified mathematical programming formulation is developed in Chapter 4 for computing the *dynamic* flexibility index (FI_d). The solution strategies of this model are outlined, and an illustrative example is also presented in this chapter.

Although a batch process may become inoperable due to instantaneous variations in some process parameters at certain instances, the cumulative effects of temporary disturbances in finite time intervals can also result in serious consequences. The mathematical programming model presented in Chapter 5 can be adopted to address this important design issue by computing the so-called *temporal* flexibility index (FI_t). Its mathematical definition and the corresponding computation procedures are also provided in this chapter. For comparison purposes, the illustrative example used here is the same as that given in Chapter 4.

Various applications of the aforementioned four flexibility indices can be found in the next four chapters. Chapter 6 presents a systematic revamping approach to enhance operational flexibility of single- and multicontaminant water networks based on the steady-state index FI_s, and Chapter 7 provides extensive case studies to demonstrate the advantages of characterizing a continuous process with the volumetric index FI_v when the steady-state flexibility index gives an overly pessimistic assessment. Both the dynamic and temporal flexibility indices are applied and compared in Chapters 8 and 9 to the solar-driven membrane distillation desalination system (SMDDS) and the hybrid power generation system, respectively. Chapter 10 contains discussions on the potential future works.

Finally, this book is written primarily for those who have a basic knowledge of optimization theory, and its presentation is oriented toward a multidisciplinary audience and thus should appeal to engineers in diverse fields with an interest in producing resilient system designs.

MATLAB® and Simulink® are registered trademarks of The MathWorks, Inc. For product information, please contact:

The MathWorks, Inc.
3 Apple Hill Drive
Natick, MA 01760-2098 USA
Tel: 508-647-7000
Fax: 508-647-7001
E-mail: info@mathworks.com
Web: www.mathworks.com

About the Authors

Chuei-Tin Chang received his BS degree in Chemical Engineering from National Taiwan University and his PhD in Chemical Engineering from Columbia University in 1976 and 1982, respectively. He worked as a process engineer for FMC Corporation (Princeton, New Jersey) from 1982 to 1985 and as an assistant professor at the Department of Chemical Engineering at the University of Nebraska (Lincoln) from 1985 to 1989. He later joined the faculty of Chemical Engineering Department of the National Cheng Kung University (Tainan, Taiwan) in 1989, became a full professor in 1993, and recently received the Distinguished Professor Award (2008–2011). He was a visiting research scholar at Northwestern University (1992), Leihigh University (1998), and Georgia Institute of Technology (2006). He is a senior member of AIChE and TwIChE. His current research interests are mainly concerned with process systems engineering (PSE), which include process integration, process safety assessment and fault diagnosis, etc. He is the author of 104 refereed papers and 90 conference papers (total number of citations: 1306; h-index: 16).

Vincentius Surya Kurnia Adi is an assistant professor in the Chemical Engineering Department, National Chung Hsing University. He received his BS in Chemical Engineering from Institut Teknologi Bandung, Indonesia, in 2005. He worked as a process engineer for PT. Arianto Darmawan until 2006, and later in 2008 received his MS in Chemical Engineering from National Cheng Kung University, Tainan, Taiwan. He finished his PhD in Chemical Engineering from National Cheng Kung University, Tainan, Taiwan, in 2013. Later on, he worked as a postdoctoral researcher in the same institution until 2015. Afterwards, he moved to Chemical Engineering Department, Imperial College London as research associate before he joined as one of the faculty members in the Chemical Engineering Department, National Chung Hsing University in 2016. His research centers on developing new insights in process synthesis and design, especially in flexibility analysis of chemical processes.

1 Introduction

Dealing with uncertainties is a practical issue encountered in designing almost every chemical process. These so-called uncertainties may arise either from random exogenous disturbances (such as those in feed qualities, product demands, environmental conditions, etc.) or from undefinable variations in the internal parameters (e.g., the heat transfer coefficients, the reaction rate constants, and other physical properties) (Lima and Georgakis, 2008; Lima et al., 2010a, b; Malcolm et al., 2007). The ability of a given system to maintain feasible operation despite uncertain deviations from the nominal conditions is usually referred to as its *operational flexibility* (Halemane and Grossmann, 1983), which is clearly a feature of critical importance that must be incorporated into the design considerations. To this end, various programming approaches to facilitate deterministic flexibility analyses have already been proposed in numerous studies (Adi and Chang, 2011; Bansal et al., 2000, 2002; Floudas et al., 2001; Grossmann and Floudas, 1987; Lima and Georgakis, 2008; Lima et al., 2010a, b; Malcolm et al., 2007; Ostrovski et al., 2002; Ostrovsky et al., 2000; Ostrovsky and Volin, 1992; Swaney and Grossmann, 1985a, b; Varvarezos et al., 1995; Volin and Ostrovskii, 2002). Traditionally, the operational flexibility of a process is ensured in an ad hoc fashion by choosing conservative operating conditions, applying empirical overdesign factors, and introducing additional or redundant units. The major drawbacks of this approach can be summarized as follows:

1. Because the interactions among units are not considered, the actual flexibility level of the entire process cannot be accurately determined.
2. Because the economic penalties of the heuristic design practices are not evaluated, their financial implications cannot be adequately assessed.

A number of mathematical programming models have been developed to facilitate quantitative flexibility analyses so as to provide the designers with the capabilities to (1) determine the performance index of any given design in relation to the expected requirements in actual operation, (2) identify the bottleneck conditions that limit the operational feasibility, and (3) compare alternative designs on an objective basis (Swaney and Grossmann, 1985a). The following four performance measures are discussed in this book.

1.1 STEADY-STATE FLEXIBILITY INDEX

This flexibility index was first defined by Swaney and Grossmann (1985a, b) for use as an unambiguous gauge of the feasible region in the parameter space. Specifically, it is associated with the maximum allowable deviations of the uncertain parameters from their nominal values, by which viable operation can be assured with proper manipulation of the control variables. Swaney and Grossmann (1985b) also showed that under certain convexity assumptions, critical points that limit feasibility and/ or flexibility must lie on the vertices of the uncertain parameter space. Grossmann and Floudas (1987) later exploited the fact that sets of active constraints are responsible for limiting the flexibility of a design and developed a mixed integer nonlinear programming (MINLP) model accordingly. Similar flexibility analysis has also been carried out in a series of subsequent studies to produce resilient grassroots and revamp designs for water networks (Chang et al., 2009; Riyanto and Chang, 2010). Because the steady-state material-and-energy balances are used as the equality constraints in the aforementioned MINLP model (Grossmann and Floudas, 1987; Ostrovski et al., 2002; Ostrovsky et al., 2000; Ostrovsky and Volin, 1992; Swaney and Grossmann, 1985a, b; Varvarezos et al., 1995; Volin and Ostrovskii, 2002), the corresponding steady-state–based index FI_s is viewed as a performance measure for the *continuous* processes (Petracci et al., 1996; Pistikopoulos and Grossmann, 1988a, b, 1989a, b).

1.2 VOLUMETRIC FLEXIBILITY INDEX

Geometrically speaking, the aforementioned index FI_s can be regarded as an aggregated measure of the orthogonal distances between the given nominal point and all faces of the biggest inscribable hypercube inside the feasible region. Hence its value may not be a truly representative indicator of the entire feasible region when the chosen nominal point is very far off from the center and/or the biggest inscribable hypercube is much smaller than the feasible region due to concavities. Lai and Hui (2008) suggested using an additional yardstick, that is, the volumetric flexibility index (denoted as FI_v), to complement the conventional steady-state flexibility analysis. Essentially, FI_v should be viewed in 3-D as the volumetric fraction of the feasible region in a cube bounded by the expected upper and lower limits of uncertain process parameters. Because the total volume of feasible region is calculated without needing to select a nominal point and/or to identify the biggest inscribable cube inside the feasible region, the magnitude of FI_v can be more closely linked to process flexibility in cases when the feasible regions are nonconvex and/or the nominal conditions are associated with near-boundary locations. Finally, note that several other alternative approaches have also been proposed to address the uncertainty issues from a stochastic viewpoint (Banerjee and Ierapetritou, 2002, 2003; Pistikopoulos and Mazzuchi, 1990; Straub and Grossmann, 1990, 1992, 1993). Because they are out of the intended scope of this book, the related studies are not reviewed here.

1.3 DYNAMIC FLEXIBILITY INDEX

As indicated by Dimitriadis and Pistikopoulos (1995), the operational flexibility of an unsteady (or batch) process should be evaluated differently. By adopting a system of differential algebraic equations (DAEs) as the model constraints, these authors developed a mathematical programming formulation for the dynamic flexibility analysis. Clearly, this practice is more rigorous than that based on the steady-state model because, even for a continuous process, the operational flexibility cannot be adequately characterized without accounting for the control dynamics. In an earlier study, Brengel and Seider (1992) advocated the need for design and control integration. The incorporation of flexibility and controllability in design consideration was later studied extensively (Aziz and Mujtaba, 2002; Bahri et al., 1997; ChaconMondragon and Himmelblau, 1996; Georgiadis and Pistikopoulos, 1999; Malcolm et al., 2007; Mohideen et al., 1996a, b). Soroush and Kravaris (1993a, b) addressed issues concerning flexible operation for batch reactors. The effects of uncertainty on the dynamic behavior of chemical processes were also studied by Walsh and Perkins (1994), with particular reference to the wastewater neutralization processes. White et al. (1996) presented an approach for the evaluation of the switchability of a proposed design, that is, its ability to perform well when moving between different operating points. Dimitriadis et al. (1997) studied the feasibility problem from the safety verification point of view. Zhou et al. (2009) utilized a similar approach to assess the operational flexibility of batch systems. This problem is considered more challenging because the nature of inherent system dynamics is dependent upon the initial conditions.

1.4   TEMPORAL FLEXIBILITY INDEX

In the aforementioned dynamic flexibility analysis, the nominal values of uncertain parameters and the anticipated positive and negative deviations in these parameters must be available in every instance over the entire time horizon of operation life. The corresponding flexibility index can be uniquely determined by such *a priori* information. However, although an ill-designed system may become inoperable due to instantaneous variations in some process parameters, the cumulative effects of temporary disturbances within finite time intervals can also result in serious operational problems. The latter scenario is usually ignored in the traditional dynamic flexibility analysis, but it is, in fact, a more likely event in practical applications. To address this issue, a new mathematical programming model has been developed by Adi and Chang (2013) for computing the corresponding performance measure, which was referred to as the *temporal flexibility index*. Realistic process improvements can be identified by this novel approach in flexibility analysis.

1.5   IMPLEMENTATION STRATEGY IN PROCESS DESIGN

In process design, optimizing a single economic index (for example, minimization of the total annual cost) is the most popular approach. However, as a general rule, the operability of a design deteriorates as the budget decreases and vice versa. Apparently, an additional quantitative measure is also needed to assess the operational performance of a practical system. The steady-state and volumetric flexibility indices (FI_s and FI_v) are applicable for characterizing the continuously operated chemical plants, whereas their dynamic and temporal counterparts (FI_d and FI_t) are meant for evaluating the unsteady or batch processes.

Because the programming model used for computing FI_s is more constrained than that for FI_v, the steady-state flexibility index of a given continuous process should be first compared with the designated target. If the desired value cannot be achieved even with a reasonable amount of extra budget, then the volumetric flexibility analysis should be performed to produce a feasible design according to a relaxed criterion and/or to identify a set of more operable nominal conditions. It should also be noted that the roles of FI_d and FI_t in designing the unsteady processes are essentially the same as those of FI_s and FI_v and, thus, the corresponding implementation strategy is identical.

REFERENCES

Adi, V.S.K., Chang, C.T., 2011. Two-tier search strategy to identify nominal operating conditions for maximum flexibility. *Industrial & Engineering Chemistry Research* 50, 10707–10716.

Adi, V.S.K., Chang, C.T., 2013. A mathematical programming formulation for temporal flexibility analysis. *Computers & Chemical Engineering* 57, 151–158.

Aziz, N., Mujtaba, I.M., 2002. Optimal operation policies in batch reactors. *Chemical Engineering Journal* 85, 313–325.

Bahri, P.A., Bandoni, J.A., Romagnoli, J.A., 1997. Integrated flexibility and controllability analysis in design of chemical processes. *Aiche Journal* 43, 997–1015.

Banerjee, I., Ierapetritou, M.G., 2002. Design optimization under parameter uncertainty for general black-box models. *Industrial & Engineering Chemistry Research* 41, 6687–6697.

Banerjee, I., Ierapetritou, M.G., 2003. Parametric process synthesis for general nonlinear models. *Computers & Chemical Engineering* 27, 1499–1512.

Bansal, V., Perkins, J.D., Pistikopoulos, E.N., 2000. Flexibility analysis and design of linear systems by parametric programming. *Aiche Journal* 46, 335–354.

Bansal, V., Perkins, J.D., Pistikopoulos, E.N., 2002. Flexibility analysis and design using a parametric programming framework. *Aiche Journal* 48, 2851–2868.

Brengel, D.D., Seider, W.D., 1992. Coordinated design and control optimization of nonlinear processes. *Computers & Chemical Engineering* 16, 861–886.

ChaconMondragon, O.L., Himmelblau, D.M., 1996. Integration of flexibility and control in process design. *Computers & Chemical Engineering* 20, 447–452.

Chang, C.T., Li, B.H., Liou, C.W., 2009. Development of a generalized mixed integer nonlinear programming model for assessing and improving the operational flexibility of water network designs. *Industrial & Engineering Chemistry Research* 48, 3496–3504.

Dimitriadis, V.D., Pistikopoulos, E.N., 1995. Flexibility analysis of dynamic-systems. *Industrial & Engineering Chemistry Research* 34, 4451–4462.

Dimitriadis, V.D., Shah, N., Pantelides, C.C., 1997. Modeling and safety verification of discrete/continuous processing systems. *Aiche Journal* 43, 1041–1059.

Floudas, C.A., Gumus, Z.H., Ierapetritou, M.G., 2001. Global optimization in design under uncertainty: Feasibility test and flexibility index problems. *Industrial & Engineering Chemistry Research* 40, 4267–4282.

Georgiadis, M.C., Pistikopoulos, E.N., 1999. An integrated framework for robust and flexible process systems. *Industrial & Engineering Chemistry Research* 38, 133–143.

Grossmann, I.E., Floudas, C.A., 1987. Active constraint strategy for flexibility analysis in chemical processes. *Computers & Chemical Engineering* 11, 675–693.

Halemane, K.P., Grossmann, I.E., 1983. Optimal process design under uncertainty. *Aiche Journal* 29, 425–433.

Lai, S.M., Hui, C.W., 2008. Process flexibility for multivariable systems. *Industrial & Engineering Chemistry Research* 47, 4170–4183.

Lima, F.V., Georgakis, C., 2008. Design of output constraints for model-based non-square controllers using interval operability. *Journal of Process Control* 18, 610–620.

Lima, F.V., Georgakis, C., Smith, J.F., Schnelle, P.D., Vinson, D.R., 2010a. Operability-based determination of feasible control constraints for several high-dimensional nonsquare industrial processes. *Aiche Journal* 56, 1249–1261.

Lima, F.V., Jia, Z., Lerapetritou, M., Georgakis, C., 2010b. Similarities and differences between the concepts of operability and flexibility: The steady-state case. *Aiche Journal* 56, 702–716.

Malcolm, A., Polan, J., Zhang, L., Ogunnaike, B.A., Linninger, A.A., 2007. Integrating systems design and control using dynamic flexibility analysis. *Aiche Journal* 53, 2048–2061.

Mohideen, M.J., Perkins, J.D., Pistikopoulos, E.N., 1996a. Optimal design of dynamic systems under uncertainty. *Aiche Journal* 42, 2251–2272.

Mohideen, M.J., Perkins, J.D., Pistikopoulos, E.N., 1996b. Optimal synthesis and design of dynamic systems under uncertainty. *Computers & Chemical Engineering* 20, S895–S900.

Ostrovski, G.M., Achenie, L.E.K., Karalapakkam, A.M., Volin, Y.M., 2002. Flexibility analysis of chemical processes: Selected global optimization sub-problems. *Optimization and Engineering* 3, 31–52.

Ostrovsky, G.M., Achenie, L.E.K., Wang, Y.P., Volin, Y.M., 2000. A new algorithm for computing process flexibility. *Industrial & Engineering Chemistry Research* 39, 2368–2377.

Ostrovsky, G.M., Volin, Y.M., 1992. Optimal process design under uncertainty. *Doklady Akademii Nauk* 325, 103–106.

Petracci, N.C., Hoch, P.M., Eliceche, A.M., 1996. Flexibility analysis of an ethylene plant. *Computers & Chemical Engineering* 20, S443–S448.

Pistikopoulos, E.N., Grossmann, I.E., 1988a. Evaluation and redesign for improving flexibility in linear-systems with infeasible nominal conditions. *Computers & Chemical Engineering* 12, 841–843.

Pistikopoulos, E.N., Grossmann, I.E., 1988b. Optimal retrofit design for improving process flexibility in linear-systems. *Computers & Chemical Engineering* 12, 719–731.

Pistikopoulos, E.N., Grossmann, I.E., 1989a. Optimal retrofit design for improving process flexibility in nonlinear-systems .1. Fixed degree of flexibility. *Computers & Chemical Engineering* 13, 1003–1016.

Pistikopoulos, E.N., Grossmann, I.E., 1989b. Optimal retrofit design for improving process flexibility in nonlinear-systems .2. Optimal level of flexibility. *Computers & Chemical Engineering* 13, 1087–1096.

Pistikopoulos, E.N., Mazzuchi, T.A., 1990. A novel flexibility analysis approach for processes with stochastic parameters. *Computers & Chemical Engineering* 14, 991–1000.

Riyanto, E., Chang, C.T., 2010. A heuristic revamp strategy to improve operational flexibility of water networks based on active constraints. *Chemical Engineering Science* 65, 2758–2770.

Soroush, M., Kravaris, C., 1993a. Optimal-design and operation of batch reactors .1. Theoretical framework. *Industrial & Engineering Chemistry Research* 32, 866–881.

Soroush, M., Kravaris, C., 1993b. Optimal-design and operation of batch reactors .2. A case-study. *Industrial & Engineering Chemistry Research* 32, 882–893.

Straub, D.A., Grossmann, I.E., 1990. Integrated stochastic metric of flexibility for systems with discrete state and continuous parameter uncertainties. *Computers & Chemical Engineering* 14, 967–985.

Straub, D.A., Grossmann, I.E., 1992. Evaluation and optimization of stochastic flexibility in multiproduct batch plants. *Computers & Chemical Engineering* 16, 69–87.

Straub, D.A., Grossmann, I.E., 1993. Design optimization of stochastic flexibility. *Computers & Chemical Engineering* 17, 339–354.

Swaney, R.E., Grossmann, I.E., 1985a. An index for operational flexibility in chemical process design .1. Formulation and theory. *Aiche Journal* 31, 621–630.

Swaney, R.E., Grossmann, I.E., 1985b. An index for operational flexibility in chemical process design .2. Computational algorithms. *Aiche Journal* 31, 631–641.

Varvarezos, D.K., Grossmann, I.E., Biegler, L.T., 1995. A sensitivity based approach for flexibility analysis and design of linear process systems. *Computers & Chemical Engineering* 19, 1301–1316.

Volin, Y.M., Ostrovskii, G.M., 2002. Flexibility analysis of complex technical systems under uncertainty. *Automation and Remote Control* 63, 1123–1136.

Walsh, S., Perkins, J., 1994. Application of integrated process and control-system design to waste-water neutralization. *Computers & Chemical Engineering* 18, S183–S187.

White, V., Perkins, J.D., Espie, D.M., 1996. Switchability analysis. *Computers & Chemical Engineering* 20, 469–474.

Zhou, H., Li, X.X., Qian, Y., Chen, Y., Kraslawski, A., 2009. Optimizing the initial conditions to improve the dynamic flexibility of batch processes. *Industrial & Engineering Chemistry Research* 48, 6321–6326.

2 Steady-State Flexibility Analysis

As mentioned previously, design and control decisions are usually made in two consecutive steps over the life cycle of a *continuous* chemical process. In the traditional design phase, the "optimal" operating conditions and the corresponding material- and energy-balance data are produced on the basis of economic considerations only. Because it is often desirable to address the operability issues at the earliest possible stage, systematic incorporation of flexibility analysis in process synthesis and design has received considerable attention in recent years (Bansal et al., 2000, 2002; Dimitriadis and Pistikopoulos, 1995; Floudas et al., 2001; Grossmann and Halemane, 1982; Swaney and Grossmann, 1985a, b). The uncertainties in flexibility analysis may be attributed to either random exogenous disturbances (such as those in feed qualities, product demands, environmental conditions, etc.) or uncharacterizable variations in the internal parameters (such as heat transfer coefficients, reaction rate constants, and other physical properties) (Lima and Georgakis, 2008; Lima et al., 2010a, b; Malcolm et al., 2007), and the ability of a chemical process to maintain feasible operation despite uncertain deviations from the nominal states was referred to as its *operational flexibility*. The so-called *flexibility index* (FI_s) was first proposed by Swaney and Grossmann (1985a, b) to provide a quantitative measure of the feasible region in the parameter space. More specifically, FI_s can be associated with the maximum allowable deviations of the uncertain parameters from their nominal values, by which feasible operation can be assured with proper manipulation of the control variables. The aforementioned authors also showed that, under certain convexity assumptions, critical points that limit feasibility and/or flexibility must lie on the vertices of the uncertain parameter space. Grossmann and Floudas (1987) later exploited the fact that active constraints are responsible for limiting the operability of a design and developed a mixed integer nonlinear programming (MINLP) model accordingly. Various approaches to facilitate the corresponding flexibility analysis have also been proposed in numerous studies (Bansal et al., 2000, 2002; Floudas et al., 2001; Grossmann and Floudas, 1987; Grossmann and Halemane, 1982; Halemane and Grossmann, 1983; Lima and Georgakis, 2008; Lima et al., 2010a, b; Malcolm et al., 2007; Ostrovski et al., 2002; Ostrovsky et al., 2000; Swaney and Grossmann, 1985a, b; Varvarezos et al., 1995; Volin and Ostrovskii, 2002). This analysis was also carried out in a series of subsequent studies to produce resilient grassroots and revamp designs (Chang et al., 2009; Riyanto and Chang, 2010). Because the steady-state material and energy balances are used as the equality constraints in the aforementioned MINLP model (Grossmann and Floudas, 1987; Ostrovski et al., 2002; Ostrovsky et al., 2000; Swaney and Grossmann, 1985a, b; Varvarezos et al., 1995; Volin and Ostrovskii, 2002), this original index can be viewed as a performance indicator of the *continuous* process under consideration

(Petracci et al., 1996; Pistikopoulos and Grossmann, 1988a, b, 1989a, b), and it is referred to as the *steady-state flexibility index* throughout this book.

2.1   MODEL FORMULATION

The steady-state flexibility index was first defined by Swaney and Grossmann (1985a, b) as a lumped indicator of the allowable variations in all uncertain parameters. The basic framework of the flexibility index model (Biegler et al., 1997) is presented in the sequel. For clarity, let us first introduce two label sets:

$$\mathbb{I} = \{i \,|\, i \text{ is the label of an equality constraint}\} \tag{2.1}$$

$$\mathbb{J} = \{j \,|\, j \text{ is the label of an inequality constraint}\} \tag{2.2}$$

The general design model of the continuous processes can be expressed accordingly as

$$h_i(\mathbf{d}, \mathbf{z}, \mathbf{x}, \theta) = 0 \qquad \forall i \in \mathbb{I} \tag{2.3}$$

$$g_j(\mathbf{d}, \mathbf{z}, \mathbf{x}, \theta) \le 0, \qquad \forall j \in \mathbb{J} \tag{2.4}$$

where h_i is the ith equality constraint in the design model (e.g., the steady-state mass or energy balance equation of a processing unit); g_j is the jth inequality constraint (e.g., a capacity limit); $\mathbf{d}$ represents a vector in which all design specifications are stored; $\mathbf{z}$ denotes the vector of adjustable control variables; $\mathbf{x}$ is the vector of state variables; and θ denotes the vector of uncertain parameters.

The following mathematical program can be utilized to determine a so-called *feasibility function* $\Psi(d, \theta)$, that is,

$$\psi(\mathbf{d}, \theta) = \min_{\mathbf{x}, \mathbf{z}} \max_{j \in \mathbb{J}} g_j(\mathbf{d}, \mathbf{z}, \mathbf{x}, \theta) \tag{2.5}$$

subject to the equality constraints given in Equation 2.3. Notice that this formulation means that for a *fixed* design defined by $\mathbf{d}$ and the *fixed* values of parameters in θ, the largest g_j ($\forall j \in \mathbb{J}$) is minimized by adjusting the control variables in $\mathbf{z}$ while keeping $h_i = 0$ ($\forall i \in \mathbb{I}$). The given system is clearly operable if $\Psi(d, \theta) \le 0$, although infeasible if otherwise (see Figure 2.1).

On the other hand, the earlier optimization problem can be defined alternatively with a different formulation by introducing an extra scalar variable, u, that is,

$$\psi(\mathbf{d}, \theta) = \min_{\mathbf{x}, \mathbf{z}, u} u \tag{2.6}$$

subject to Equation 2.3 and

$$g_j(\mathbf{d}, \mathbf{z}, \mathbf{x}, \theta) \le u \qquad \forall j \in \mathbb{J} \tag{2.7}$$

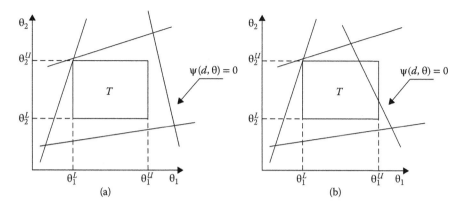

FIGURE 2.1 Feasible and infeasible designs in the parameter space: (a) is feasible because $\psi(d,\theta) \leq 0$; (b) is infeasible because $\psi(d,\theta) > 0$.

Notice also that if $\Psi(d,\theta) = 0$, then at least one of the inequality constraints should be active, that is, $g_j = 0$ ($\exists\, j \in \mathbb{J}$).

Because the aforementioned feasibility function is evaluated according to a deterministic model with *constant* θ, it is necessary to perform the feasibility check on a more comprehensive basis by considering all possible values of the uncertain parameters. To this end, let us first define a permissible hypercube **T** in the parameter space, that is,

$$\mathbf{T} = \left\{ \theta \mid \theta^N - \Delta\theta^- \leq \theta \leq \theta^N + \Delta\theta^+ \right\} \tag{2.8}$$

where θ^N denotes a vector of the given nominal parameter values and $\Delta\theta^+$ and $\Delta\theta^-$ represent vectors of the expected deviations in the positive and negative directions, respectively. Hence, an additional optimization problem can be formulated to facilitate this more rigorous test:

$$\chi(\mathbf{d}) = \max_{\theta \in \mathbf{T}} \psi(\mathbf{d},\theta) \tag{2.9}$$

where $\chi(d)$ denotes the feasibility function of a *fixed* design defined by **d** over **T**. The given system should therefore be feasible if $\chi(d) \leq 0$ and infeasible if otherwise (Figure 2.2).

To provide a unified measure of the maximum tolerable range of variation in every uncertain parameter (Swaney and Grossmann, 1985a, b), the permissible hypercube **T** is expanded/contracted with another scalar variable, δ:

$$\mathbf{T}(\delta) = \left\{ \theta \mid \theta^N - \delta\Delta\theta^- \leq \theta \leq \theta^N + \delta\Delta\theta^+ \right\} \tag{2.10}$$

where δ can be determined by solving the flexibility index model given as:

$$FI_s = \max \delta \tag{2.11}$$

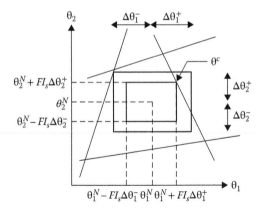

FIGURE 2.2   Geometrical interpretation of the steady-state flexibility index.

subject to

$$\chi(\mathbf{d}) \leq 0 \tag{2.12}$$

Note that the maximized objective value FI_s is the steady-state *flexibility index*, which represents the largest value of δ that guarantees $g_j \leq 0$ ($\forall j \in \mathbb{J}$), that is, $\chi(\mathbf{d}) \leq 0$, in the parameter hypercube. Note also that $\delta \geq 1$ essentially implies that the system is feasible under the original constraints of Equation 2.8.

2.2   NUMERICAL SOLUTION STRATEGIES

Several effective strategies are available for solving the optimization problem defined by Equations 2.11 and 2.12. Two of them, that is, the active set method and the vertex method, are described in the sequel.

2.2.1   ACTIVE SET METHOD

Solving the flexibility index model is, in general, very tough because Equations 2.11 and 2.12 represent a nonlinear, nondifferentiable, multilevel optimization problem. Grossmann and Floudas (1987) developed a solution strategy based on the Karush-Kuhn-Tucker (KKT) conditions of the optimization problem for computing the function $\Psi(\mathbf{d}, \theta)$, that is, Equations 2.3, 2.6, and 2.7. To be able to apply these conditions, the aforementioned flexibility index model is first reformulated by imposing an extra equality constraint that forces the feasibility function to be zero, that is,

$$FI_s = \min \delta \tag{2.13}$$

subject to

$$\psi(\mathbf{d}, \theta) = 0 \tag{2.14}$$

Notice that the original maximization problem—that is, Equations 2.11 and 2.12—is now replaced with the present *minimization* problem. This is because if the chosen value of δ is not the smallest, at least one inequality constraint must be violated, that is, $g_j > 0$ ($\exists j \in \mathbb{J}$). Because Equations 2.3 and 2.4 are inherently satisfied in the optimization problem defined by Equations 2.3, 2.6, and 2.7, the corresponding KTT conditions should be applicable. Consequently, the flexibility evaluation problem in Equations 2.13 and 2.14 can be reformulated more explicitly as an MINLP model (Grossmann and Floudas, 1987):

$$FI_s = \min_{\delta,\mu_i,\lambda_j,s_j,y_j,x_i,z_c,\theta_n} \delta \tag{2.15}$$

subject to the equality constraints in Equation 2.3 and also those presented as follows:

$$g_j(\mathbf{d},\mathbf{z},\mathbf{x},\boldsymbol{\theta})+s_j = 0, \quad j \in \mathbb{J} \tag{2.16}$$

$$\sum_{i\in\mathbb{I}}\mu_i\frac{\partial h_i}{\partial z}+\sum_{j\in\mathbb{J}}\lambda_j\frac{\partial g_j}{\partial z}=0 \tag{2.17}$$

$$\sum_{i\in\mathbb{I}}\mu_i\frac{\partial h_i}{\partial \mathbf{x}}+\sum_{j\in\mathbb{J}}\lambda_j\frac{\partial g_j}{\partial \mathbf{x}}=0 \tag{2.18}$$

$$\sum_{j\in\mathbb{J}}\lambda_j = 1 \tag{2.19}$$

$$\lambda_j - y_j \leq 0, \quad j \in \mathbb{J} \tag{2.20}$$

$$s_j - Q(1-y_j) \leq 0, \quad j \in \mathbb{J} \tag{2.21}$$

$$\sum_{j\in\mathbb{J}}y_j = n_z + 1 \tag{2.22}$$

$$\theta^N - \delta\Delta\theta^- \leq \theta \leq \theta^N + \delta\Delta\theta^+ \tag{2.23}$$

$$y_j = \{0,1\},\lambda_j \geq 0, s_j \geq 0, \quad j \in \mathbb{J} \tag{2.24}$$

$$\delta \geq 0 \tag{2.25}$$

where, s_j is the slack variable for the jth inequality constraint; Q denotes a large enough positive number to be used as the upper bound of s_j; μ_i denotes the Lagrange multiplier of equality constraint h_i; λ_j is the Lagrange multiplier of inequality constraint g_j; y_j is the binary variable reflecting whether the corresponding inequality constraint is active, that is, $g_j = 0$ if $y_j = 1$, whereas $g_j < 0$ if $y_j = 0$; and n_z is the total number of independent control variables.

2.2.2 Vertex Method

The primary difficulty in applying the active set method can be clearly attributed to the computation effort involved in solving the aforementioned MINLP model. The optimization procedure for Equations 2.11 and 2.12 can be greatly simplified under the assumption that the optimal solution is always associated with one of the vertices of the feasible hypercube in the parameter space (Halemane and Grossmann, 1983). Let $\Delta\theta^k$ $(\forall k \in V)$ denote a vector pointing from the nominal point θ^N to the kth vertex and V is the set of all vertices. Then it is possible to determine the largest possible value of δ along a specific vertex direction, that is, $\Delta\theta^k$, by solving the following programming model:

$$\delta^k = \max_{\mathbf{x},\mathbf{z},\delta} \delta \qquad (2.26)$$

subject to Equations 2.3, 2.4, and

$$\theta = \theta^N + \delta\Delta\theta^k \qquad (2.27)$$

Among all resulting parameter hypercubes, that is, $\mathbf{T}(\delta^k)$ and $k \in V$, it is clear that only the smallest one can be totally inscribed within the feasible region defined by Equations 2.3 and 2.4. Hence,

$$FI_s = \min_{k \in V}\{\delta^k\} \qquad (2.28)$$

Thus the following simple procedure applies:

Step 1: Solve the optimization problem described by Equations 2.3, 2.4, 2.26, and 2.27 for each vertex $k \in V$.
Step 2: Select FI_s according to Equation 2.28.

Swaney and Grossmann (1985a, b) showed that only under certain convexity conditions is the optimal solution guaranteed to be associated with one of the vertices. However, even when these conditions are not met, it can often be found that this approach is still applicable in practice. Note also that the vertex method may be computationally demanding as the number of uncertain parameters increases. For example, $1,024 \ (= 2^{10})$ optimization runs are needed for 10 uncertain parameters, and if the number of parameters is raised to 20, the computation load for the required $1,048,576 \ (= 2^{20})$ runs can be quite overwhelming. However, in certain realistic applications, a significant portion of theses runs may be omitted on the basis of physical insights (Li and Chang, 2011).

2.3 ILLUSTRATIVE EXAMPLES

The implementation steps of steady-state flexibility analysis are illustrated here with two simple examples.

Example 2.1

Let us first consider the water network shown in Figure 2.3 (Chang et al., 2009). Note that there are one water source (W), one wastewater sink (S), and two water-using units, that is, $U1$ and $U2$, in this system. The supply rate of the freshwater source (fww) is set to be 430 ton/hr, and its contaminant concentration (Cw) is 20 ppm. The nominal operating conditions of the two water-using units are shown in Table 2.1.

Let us also assume that the upper limits of inlet and outlet concentrations of the two water-using units, that is, $C_{in,unit}^{max}$ and $C_{out,unit}^{max}$, are affected by the ambient temperature and thus can be regarded as uncertain parameters, specifically

$$C_{in,unit}^{max} = \theta \overline{C}_{in,unit}^{max} \tag{2.29}$$

$$C_{out,unit}^{max} = \theta \overline{C}_{out,unit}^{max} \tag{2.30}$$

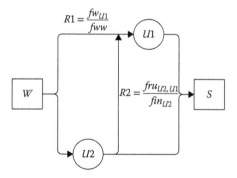

FIGURE 2.3 The water network studied in Example 2.1. (Reprinted with permission from [Chang et al., 2009, 3496–3504]. Copyright [2009] American Chemical Society.)

TABLE 2.1
Nominal Operating Conditions of Water-Using Units in Example 2.1

Water Using Unit	$\overline{C}_{in}^{max}$ (ppm)	$\overline{C}_{out}^{max}$ (ppm)	Mass Load (kg/hr)
$U1$	70	170	20
$U2$	N/A	120	30

Source: Reprinted with permission from [Chang et al., 2009, 3496–3504]. Copyright [2009] **American Chemical Society**.

where the subscript *unit* can be replaced with either $U1$ or $U2$; $\bar{C}_{in,unit}^{max}$ and $\bar{C}_{out,unit}^{max}$, respectively, denote the nominal values of corresponding concentration limits; and θ is an uncertain multiplier.

The following two flow ratios are treated as the control variables in this example:

$$R1 = \frac{fw_{U1}}{fww} \tag{2.31}$$

$$R2 = \frac{fru_{U2,U1}}{fin_{U2}} \tag{2.32}$$

where fww is the supply rate from source W (ton/hr); fw_{U1} is the flow rate from W to $U1$ (ton/hr); fin_{U2} is the inlet flow rate of $U2$ (ton/hr); and $fru_{U2,U1}$ is the flow rate from $U2$ to $U1$ (ton/hr). The equality and inequality constraints can thus be expressed as

$$R1(Cw - Cin_{U1}) + (1 - R1)R2(Cout_{U2} - Cin_{U1}) = 0 \tag{2.33}$$

$$fww\left[R1(Cw - Cout_{U1}) + (1 - R1)R2(Cout_{U2} - Cout_{U1})\right] + ml_{U1} = 0 \tag{2.34}$$

$$(1 - R1)fww(Cw - Cout_{U2}) + ml_{U2} = 0 \tag{2.35}$$

$$Cin_{U1} \leq \theta\bar{C}_{in,U1}^{max} \tag{2.36}$$

$$Cout_{U1} \leq \theta\bar{C}_{out,U1}^{max} \tag{2.37}$$

$$Cout_{U2} \leq \theta\bar{C}_{out,U2}^{max} \tag{2.38}$$

where ml_{U1} is the mass load of $U1$ (kg/hr); ml_{U2} is the mass load of $U2$ (kg/hr); Cin_{U1} is the inlet concentration of $U1$ (ppm); $Cout_{U1}$ is the outlet concentration of $U1$ (ppm); $Cout_{U2}$ is the outlet concentration of $U2$ (ppm); $\bar{C}_{in,U1}^{max}$ is the nominal value of maximum inlet concentration of $U1$ (ppm); $\bar{C}_{in,U1}^{max}$ is the nominal value of maximum outlet concentration of $U1$ (ppm); and $\bar{C}_{out,U2}^{max}$ is the nominal maximum outlet concentration of $U2$ (ppm). Finally, it is assumed that the negative and positive deviations of uncertain parameters is 0.05 and 0.04, respectively, that is,

$$1 - 0.04\delta \leq \theta \leq 1 + 0.05\delta \tag{2.39}$$

The steady-state flexibility index is first computed here with the active set method. Specifically, the flexibility index model can be formulated according to Equations 2.15 through 2.25:

$$FI_s = \min_{\delta,\mu,\lambda,s,y,\mathbf{x},\mathbf{z}} \delta \tag{2.40}$$

subject to

$$R1(Cw - Cin_{U1}) + (1-R1)R2(Cout_{U2} - Cin_{U1}) = 0 \tag{2.41}$$

$$fww\left[R1(Cw - Cout_{U1}) + (1-R1)R2(Cout_{U2} - Cout_{U1})\right] + ml_{U1} = 0 \tag{2.42}$$

$$(1-R1)fww(Cw - Cout_{U2}) + ml_{U2} = 0 \tag{2.43}$$

$$Cin_{U1} - \overline{C}^{max}_{in,U1}\theta + S_{in,U1} = 0 \tag{2.44}$$

$$Cout_{U1} - \overline{C}^{max}_{out,U1}\theta + S_{out,U1} = 0 \tag{2.45}$$

$$Cout_{U2} - \overline{C}^{max}_{out,U2}\theta + S_{out,U2} = 0 \tag{2.46}$$

$$\lambda_{in,U1} + \lambda_{out,U1} + \lambda_{out,U2} = 1 \tag{2.47}$$

$$\frac{\partial L}{\partial Cin_{U1}} \Rightarrow -\mu_{cin1}\left[R1 + (1-R1)R2\right] + \lambda_{in,U1} = 0 \tag{2.48}$$

$$\frac{\partial L}{\partial Cout_{U1}} \Rightarrow -\mu_{cu1}\left[R1fww + fww(1-R1)R2\right] + \lambda_{out,U1} = 0 \tag{2.49}$$

$$\frac{\partial L}{\partial Cout_{U2}} \Rightarrow \mu_{cu1}fww(1-R1)R2 - \mu_{cu2}(1-R1)fww + \mu_{cu1}(1-R1)R2 + \lambda_{out,U2} = 0 \tag{2.50}$$

$$\frac{\partial L}{\partial R1} \Rightarrow \mu_{cin1}\left[Cw - Cin_{U1} - R2(Cout_{U2} - Cin_{U1})\right]$$
$$+ \mu_{cu1}fww\left[Cw - Cout_{U1} - R2(Cout_{U2} - Cout_{U1})\right] \tag{2.51}$$
$$- \mu_{cu2}fww(Cw - Cout_{U2}) = 0$$

$$\frac{\partial L}{\partial R2} \Rightarrow \mu_{cin1}(Cout_{U2} - Cin_{U1}) + \mu_{cu1}fww(Cout_{U2} - Cout_{U1}) = 0 \tag{2.52}$$

$$S_{\text{in},U1} - Q(1 - y_{\text{in},U1}) \le 0 \tag{2.53}$$

$$S_{\text{out},U1} - Q(1 - y_{\text{out},U1}) \le 0 \tag{2.54}$$

$$S_{\text{out},U2} - Q(1 - y_{\text{out},U2}) \le 0 \tag{2.55}$$

$$y_{\text{in},U1} + y_{\text{out},U1} + y_{\text{out},U2} \le n_z + 1 \tag{2.56}$$

$$\lambda_{\text{in},U1}, \lambda_{\text{out},U1}, \lambda_{\text{out},U2} \ge 0 \tag{2.57}$$

$$S_{\text{in},U1}, S_{\text{out},U1}, S_{\text{out},U2} \ge 0 \tag{2.58}$$

$$1 - 0.04\delta \le \theta \le 1 + 0.05\delta \tag{2.59}$$

$$0 \le R1, \ R2 \le 1 \tag{2.60}$$

where μ_{cin1}, μ_{cu1}, μ_{cu2}, $\lambda_{\text{in},U1}$, $\lambda_{\text{out},U1}$, and $\lambda_{\text{out},U2}$ are the Lagrange multipliers for Equations 2.33 through 2.38, respectively; $S_{\text{in},U1}$, $S_{\text{out},U1}$, and $S_{\text{out},U2}$ are the slack variables for Equations 2.36 through 2.38; $y_{\text{in},U1}$, $y_{\text{out},U1}$, and $y_{\text{out},U2}$ are the corresponding binary variables and $n_z = 5 - 3 = 2$; and L is the augmented objective function, that is, $L = u + \mu^T g + \lambda^T h$. Equations 2.41 through 2.43 are essentially the same as Equations 2.33 through 2.35, whereas Equations 2.44 through 2.46 are the modified versions of Equations 2.36 through 2.38 with the slack variables. Equation 2.47 is based on Equation 2.19. Equations 2.48 through 2.52 are the partial derivatives of L with respect to Cin_{U1}, $Cout_{U1}$, $Cout_{U2}$, $R1$, and $R2$, respectively. Equations 2.53 through 2.55 are derived based on Equations 2.20 and 2.21. Equation 2.56 is essentially based on Equation 2.22.

By solving Equations 2.40 through 2.60, the steady-state flexibility index was found to be $\delta = 1.453$. This value implies that the network design is operable or has enough flexibility to counteract all possible disturbances by adjusting the control variables, that is, the flow ratios $R1$ and $R2$. The corresponding critical active constraints are those given in Equations 2.44 through 2.46; that is, the upper limits of inlet concentration of $U1$ ($\bar{C}_{\text{in},U1}^{\max}$), the outlet concentration of $U1$ ($\bar{C}_{\text{out},U1}^{\max}$), and the outlet concentration of $U2$ ($\bar{C}_{\text{out},U2}^{\max}$) are all reached in the worst-case scenario.

It should also be noted that due to the special model structure for water network designs, Li and Chang (2011) suggested that a shortcut version of the vertex method can be applied by checking only *a single corner* of the parameter space. This critical point can be identified on the basis of physical insights (Chang et al., 2009) and, specifically, should be associated with

- The *upper bounds* of the mass loads of water using units and the pollutant concentrations at the primary and secondary sources
- The *lower bounds* of the removal ratios of wastewater treatment units, the allowed maximum inlet and outlet pollutant concentrations of water-using units, and the allowed maximum inlet pollutant concentration of wastewater treatment units

The flexibility index of a water network can thus be determined according to this most constrained point alone. In the present example, its location in the parameter space is apparently corresponding to the lower bounds of the allowed maximum inlet and outlet pollutant concentrations of the two water-using units. In other words, Equation 2.27 can be written as $\theta = 1 - 0.04\delta$, and the resulting flexibility index was found to be the same as before ($\delta = 1.453$).

Example 2.2

Let us next consider the heat exchanger network (HEN) given in Figure 2.4 (Kuo, 2015). Seven exchangers and one heater are embedded in this HEN to facilitate heat transfers from four hot process streams (H1 – H4) and one hot utility to three cold process streams (C1 – C3). In this example, the supply temperatures of all seven process streams are treated as uncertain parameters, and their nominal values are $T_{H1.in}^N = 400\ K$, $T_{H2.in}^N = 450\ K$, $T_{H3.in}^N = 400\ K$, $T_{H4.in}^N = 430\ K$, $T_{C1.in}^N = 310\ K$, $T_{C2.in}^N = 290\ K$, and $T_{C3.in}^N = 285\ K$. The expected maximum positive and negative deviations of each supply temperature are both set at 10 K. The utility consumption rate in heater H, that is, Q, is selected as the control variable of the given system.

The following eight equality constraints can be constructed according to the basic principle of energy balance around every unit in the given HEN:

$$F_{H1}\left(T_{H1}^{in} - T_{H1}^{*1}\right) = F_{C1}\left(T_{C1}^{*1} - T_{C1}^{in}\right) \tag{2.61}$$

$$F_{H2}\left(T_{H2}^{in} - T_{H2}^{out}\right) = F_{C1}\left(T_{C1}^{out} - T_{C1}^{*1}\right) \tag{2.62}$$

$$F_{H4}\left(T_{H4}^{*5} - T_{H4}^{*3}\right) = F_{C2}\left(T_{C2}^{*3} - T_{C2}^{in}\right) \tag{2.63}$$

$$F_{H1}(T_{H1}^{*1} - T_{H1}^{out}) = F_{C2}(T_{C2}^{*4} - T_{C2}^{*3}) \tag{2.64}$$

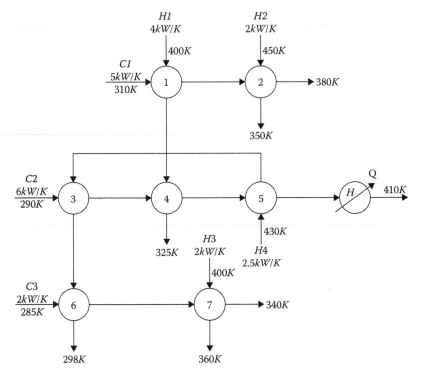

FIGURE 2.4   Heat exchanger network studied in Example 2.2. (Reprinted with permission from Kuo 2015. Copyright 2015 National Cheng Kung University Library.)

$$F_{H4}(T_{H4}^{in} - T_{H4}^{*5}) = F_{C2}(T_{C2}^{*5} - T_{C2}^{*4}) \qquad (2.65)$$

$$F_{H4}(T_{H4}^{*3} - T_{H4}^{out}) = F_{C3}(T_{C3}^{*6} - T_{C3}^{in}) \qquad (2.66)$$

$$F_{H3}(T_{H3}^{in} - T_{H3}^{out}) = F_{C3}(T_{C3}^{out} - T_{C3}^{*6}) \qquad (2.67)$$

$$Q = F_{C2}(T_{C2}^{out} - T_{C2}^{*5}) \qquad (2.68)$$

where F denotes the heat capacity flow rate of a process stream; T^{in} and T^{out} represent the supply and target temperatures of a process stream, respectively; and T^{*i} denotes the process stream temperature at the outlet of the exchanger i. Note that the subscripts of the aforementioned heat capacity flow rates and temperatures are used to represent the corresponding process streams.

The inequality constraints are primarily concerned with the process temperatures, specifically

- The temperature of the hot stream at the outlet of a heat exchanger should be higher than or equal to that of the cold stream at the inlet, whereas the inlet temperature of the hot stream should also be greater than or equal to the outlet temperature of the cold stream.
- The temperature of the hot stream at the inlet of a heat exchanger should be higher than or equal to that of the same stream at the outlet, whereas the inlet temperature of the cold stream should be lower than or equal to the outlet temperature of the same stream.

All corresponding inequalities are listed as follows:

$$T_{C1}^{in} \leq T_{H1}^{*1} \tag{2.69}$$

$$T_{C1}^{in} \leq T_{C1}^{*1} \tag{2.70}$$

$$T_{C1}^{*1} \leq T_{H2}^{out} \tag{2.71}$$

$$T_{C2}^{in} \leq T_{H4}^{*3} \tag{2.72}$$

$$T_{C2}^{in} \leq T_{C2}^{*3} \tag{2.73}$$

$$T_{C2}^{*3} \leq T_{H1}^{out} \tag{2.74}$$

$$T_{C2}^{*3} \leq T_{C2}^{*4} \tag{2.75}$$

$$T_{C2}^{*3} \leq T_{H4}^{*5} \tag{2.76}$$

$$T_{C2}^{*4} \leq T_{C2}^{*5} \tag{2.77}$$

$$T_{C2}^{*4} \leq T_{H1}^{*1} \tag{2.78}$$

$$T_{C2}^{*4} \leq T_{H4}^{*5} \tag{2.79}$$

$$T_{C2}^{*5} \leq T_{C2}^{\text{out}} \tag{2.80}$$

$$T_{C3}^{\text{in}} \leq T_{C3}^{*6} \tag{2.81}$$

$$T_{C3}^{\text{in}} \leq T_{H4}^{\text{out}} \tag{2.82}$$

$$T_{C3}^{*6} \leq T_{H3}^{\text{out}} \tag{2.83}$$

$$T_{C3}^{*6} \leq T_{H4}^{*3} \tag{2.84}$$

$$T_{H4}^{\text{out}} \leq T_{H4}^{*3} \tag{2.85}$$

$$T_{H4}^{*3} \leq T_{H4}^{*5} \tag{2.86}$$

$$T_{H4}^{*5} \leq T_{H4}^{\text{in}} \tag{2.87}$$

By using the active set and vertex methods, one can produce the following results:

- The former approach yields a flexibility index of 0.75. The inequality constraints in Equations 2.79 and 2.81 are activated in this case.
- The second approach calls for checking 2^7 vertices because there are seven uncertain parameters. The smallest δ^k among all corresponding candidates is also 0.75.

Other than the fact that both approaches may be adopted to produce the same correct results, in the sequel their pros and cons in evaluating the steady-state flexibility index are further compared:

- The active set method only requires solving a single optimization problem. However, due to the need to incorporate the KKT conditions, additional Lagrange multipliers, slack variables, and binary variables must be introduced into the model formulation. Consequently, the required optimization procedure consists of mostly the steps for solving a complex MINLP with scores of constraints and variables. Note that the convergence of the numerical solution process for this MINLP model is not guaranteed.

- A total of 128 (2^7) nonlinear programming (NLP) models are solved to determine the value of FI_s in the present example. The required computation load can become overwhelming in cases where there are more uncertain parameters. However, if a subset of the corresponding vertices can be ignored based on physical insights, one would expect the effort to obtain optimum solution may be significantly reduced. In the HEN presented in Figure 2.4, the following observations can be made:

  1. Let us first consider exchanger 6. Although the exit temperature of hot stream H4 from the exchanger must be fixed at 298 K (which should be larger than T_{C3}^{*6}), the inlet temperature of cold stream C3 is uncertain. If T_{C3}^{in} deviates to the biggest extent from its nominal value of 285 K in the positive direction, then the inequality constraint in Equation 2.81 may be violated. Thus, in applying the vertex method, this uncertain temperature should be fixed at 295 ($= 285 + 10$) K and, as a result, the number of vertices that require checking can be reduced from 2^7 to 2^6.

  2. Let us next consider exchanger 7. Note that after fixing T_{C3}^{in} at its upper bound (295 K) according to the rationale outlined earlier, the difference between the target and supply temperatures of cold stream C3 should be lowered to 45 K, and the corresponding heat consumption rate is 90 kW. Note also that the largest possible heat supply rate from hot stream H3 to exchanger 7 is 100 kW (by setting T_{H3}^{in} at its maximum value of 410 K). Thus, fixing T_{H3}^{in} at its upper limit can further reduce the vertex number to 2^5, which represents one-fourth of the original computation load.

2.4 SELECTION OF NOMINAL CONDITIONS

As mentioned before, the term *flexibility* is regarded as the capability of a system to function adequately under various sources of uncertainties. It has been recognized that these uncertainties might be due to inaccuracies in the estimates of model parameters for design calculations (such as heat transfer coefficients, reaction rate constants, and other physical properties) or external disturbances in process conditions during actual operations (such as the qualities and flow rates of feed streams). The latter conditions often fluctuate online within some statistically determinable ranges, whereas their nominal values can usually be stipulated and adjusted offline.

Traditionally, design and control decisions are made in sequential stages over the life cycle of a chemical plant. In the design phase, the "optimal" operating conditions and the corresponding material and energy balance data are determined mainly by economic considerations. In the subsequent step, the control systems are configured to maintain the critical process conditions at the fixed nominal values. Because it is often desirable to address the operability issues at the earliest possible stage, the systematic incorporation of flexibility analysis in process synthesis and design has received considerable attention in recent years. The potential benefits of manipulating the nominal values of uncertain parameters are twofold. First, the operational flexibility of a given chemical plant could be enhanced without extra capital investments. Also, the operating cost of a system with higher flexibility is arguably lower because the system can cope with more extreme abnormal conditions without shutdown. Therefore, there are legitimate incentives to develop an effective optimization strategy for selecting the best nominal conditions so as to maximize the flexibility index.

2.4.1 MODIFIED PROBLEM DEFINITION

Let us assume that for a given system, the conventional flexibility index model (see Section 2.1) is available and, also, the nominal conditions in θ^N can be divided into two different types, that is,

$$\theta^N = \begin{bmatrix} \theta_I^N \\ \theta_{II}^N \end{bmatrix} \tag{2.88}$$

where

1. θ_I^N is alterable offline with existing equipment (e.g., feed quality and flow rate, removal ratio).
2. θ_{II}^N is unalterable (e.g., heat transfer coefficients, reaction rate constants, physical properties).

Based on the initial estimates of the nominal values of both types of parameters, the modified search results should include (1) the maximum value of FI_s for the system considered and (2) the *optimal* nominal values of θ_I^N.

Notice that the conventional flexibility index model only deals with *fixed* nominal parameter values. To find the optimal FI_s by varying the nominal operating conditions in θ_I^N, a multilevel optimization procedure is needed, that is,

$$FI_s^{\max} = \max_{\theta_I^N} FI_s\left(\theta_I^N, \theta_{II}^N\right) \tag{2.89}$$

Almost all constraints of this optimization problem should be the same as those used in the original flexibility index model—that is, Equations 2.1 through 2.5, 2.9, 2.11, and 2.12—whereas Equation 2.10 should be replaced by Equation 2.88 and

$$\theta_I^{\min} \le \theta_I^N - \delta\Delta\theta_I^- \le \theta_I \le \theta_I^N + \delta\Delta\theta_I^+ \le \theta_I^{\max}$$

$$\theta_{II}^N - \delta\Delta\theta_{II}^- \le \theta_{II} \le \theta_{II}^N + \delta\Delta\theta_{II}^+ \tag{2.90}$$

where $\theta_I^{\min}$ and $\theta_I^{\max}$, respectively, represent the lower and upper limits of acceptable values of the type-I parameters. These limits are necessary because θ_I^N is primarily viewed as a vector of adjustable *decision variables* and, therefore, the corresponding parameter intervals, that is, $\theta_I^N - \delta\Delta\theta_I^- \le \theta_I \le \theta_I^N + \delta\Delta\theta_I^+$, are no longer bounded. Notice that $\Delta\theta_I^-$ and $\Delta\theta_I^+$ are used in the proposed model to characterize the ranges of *statistically* uncertain parameters, but θ_I^{min} and θ_I^{max} are needed for setting the lower and upper bounds, which are economically feasible and/or *physically* realizable.

2.4.2 TWO-TIER SEARCH STRATEGY

The optimization problem defined earlier can be tackled hierarchically on two levels. First the MINLP described by Equations 2.3 and 2.15 through 2.22, 2.24, 2.25, and 2.90 is solved on the lower level to minimize δ based on a set of fixed nominal parameters in θ^N, whereas the upper level of the maximum value of the flexibility index is determined by adjusting the nominal operating conditions in θ_I^N according to a direct search strategy. The main reason for selecting a direct search approach in this framework is due to the need to simplify problem formulation by avoiding the use of gradients. One of the most popular multi-agent direct search strategies is the so-called genetic algorithm (GA)—see Goldberg (1989) and Holland (1975, 1992). There are two other closely related optimization methods, namely, differential evolution (DE) and particle swarm optimization (PSO). DE is conceptually similar to GA in its use of evolutionary operators to guide the search toward an optimum, but it was specifically developed for real-valued search spaces from the start. PSO was originally intended as a model for the social behavior in a flock of birds, but the algorithm was later simplified for solving optimization problems.

Earlier studies showed that DE fared best on some benchmark problems (Pedersen, 2008); therefore, it is chosen for addressing the illustrative example described in the next subsection. Specifically, this two-tier search strategy (Adi and Chang, 2011) can be concisely depicted with the flowchart in Figure 2.5. A brief explanation of each step is provided as follows:

1. *Assemble parameter values and model constraints.* Obtain all parameter values in the proposed model, including the initial estimates of θ_I^N. Formulate the objective function and all model constraints of the lower-level optimization problem based on Equations 2.3 and 2.15 through 2.22, 2.24, 2.25, and 2.90.
2. *Construct computer codes for solving the generalized flexibility index model.* Build a GAMS code in a script file according to the model formulation assembled in step 1. The model parameters in θ_I^N are allowed to be varied via the MATLAB-GAMS interface (Ferris, 2005).
3. *Generate new agents based on the reference position.* Use the best candidate as a reference to create new agents with the DE optimizer, for example, see Storn and Price (1997).
4. *Compute flexibility indices according to the reference position and the positions of new agents.* Execute the aforementioned GAMS code repeatedly with the BARON solver to determine a collection of flexibility indices using the reference parameter values in θ_I^N and those specified in the *NP* new agents.

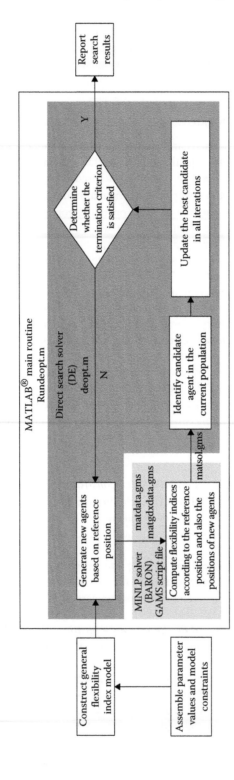

FIGURE 2.5 Two-tier search procedure. (Reprinted with permission from [Adi and Chang, 2011, 10707–10716. Copyright (2011)] American Chemical Society.)

5. *Identify the candidate agent in the current population.* The agent yielding the highest FI_s value is picked from the population in the present iteration.

6. *Update the best candidate in all iterations.* If the FI_s value of the current candidate is larger than that of the best candidate in the previous iteration, the current candidate is adopted to replace the old one. Otherwise, it is discarded.

7. *Determine whether the termination criterion is satisfied.* The iteration process is terminated when an assigned iteration number is reached or an adequate level of fitness is achieved, that is,

$$\frac{\theta_{I,k}^N - \theta_{I,k+1}^N}{\theta_{I,k}^N} \leq \varepsilon \tag{2.91}$$

where ε is a vector of error bounds and k denotes the iteration number.

8. *Report search results.* The search results are mainly the optimal θ_I^N and the corresponding $FI_s^{\max}$.

Example 2.3

The underlying water network considered here consists of one primary source $W1$, one secondary source $W2$, one sink $S1$, two water-using units $U1$ and $U2$, and a wastewater treatment unit $T1$ (see Figure 2.6). This structure was studied by Riyanto and Chang (2010), and the corresponding model parameters can be found in Table 2.2. Three uncertain parameters are considered in the present example, that is, the upper concentration limit of $W2$ (C_{W2}) and the mass loads of $U1$ and $U2$ (M_{u1} and M_{u2}). To characterize uncertainties more consistently, these parameters are normalized in the design model, that is,

$$C_{w2} = \bar{C}_{w2}\theta_{C_{w2}} \tag{2.92}$$

$$M_{u1} = \bar{M}_{u1}\theta_{M_{u1}} \tag{2.93}$$

$$M_{u2} = \bar{M}_{u2}\theta_{M_{u2}} \tag{2.94}$$

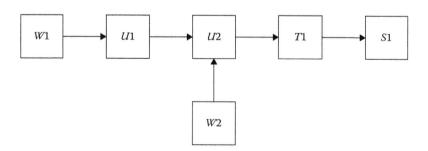

FIGURE 2.6   The basic structure of the water network in Example 2.3. (Reprinted with permission from Riyanto 2009. Copyright 2009 National Cheng Kung University Library.)

TABLE 2.2
Model Parameters Used in Water Network Problem in Example 2.3

Parameters			Values
$W1$ maximum flow rate	F_{w1}^{U}	(t/h)	35.000
$W2$ flow rate	F_{w2}	(t/h)	30.000
$W1$ concentration	C_{w1}	(ppm)	0.100
$W2$ concentration	$\bar{C}_{w2}$	(ppm)	100.000
$U1$ maximum inlet concentration	CI_{u1}^{U}	(ppm)	1
$U2$ maximum inlet concentration	CI_{u2}^{U}	(ppm)	80
$T1$ maximum inlet concentration	CI_{t1}^{U}	(ppm)	185
$S1$ maximum concentration	C_{s1}^{U}	(ppm)	30
$U1$ maximum outlet concentration	CO_{u1}^{U}	(ppm)	101
$U2$ maximum outlet concentration	CO_{u2}^{U}	(ppm)	240
$T1$ maximum flow rate	F_{t1}^{U}	(t/h)	125
$U1$ mass load	$\bar{M}_{u1}$	(kg/h)	2
$U1$ maximum tolerable mass load	M_{u1}^{U}	(kg/h)	4
$U2$ mass load	$\bar{M}_{u2}$	(kg/h)	5
$U2$ maximum tolerable mass load	M_{u1}^{U}	(kg/h)	8
$T1$ removal ratio	RR_{t1}		0.9

Source: Reprinted with permission from Riyanto (2009). Copyright 2009 National Cheng Kung University Library.

where $\bar{C}_{w2}$, $\bar{M}_{u1}$, and $\bar{M}_{u2}$ denote the reference parameter values, and $\theta_{C_{w2}}$, $\theta_{M_{u1}}$, and $\theta_{M_{u2}}$ are the corresponding *uncertain multipliers*. The uncertain multipliers are assumed to be located within the parameter space defined by Equation 2.8, in which all nominal levels are one, that is,

$$\theta_{C_{w2}}^{N} = \theta_{M_{u1}}^{N} = \theta_{M_{u2}}^{N} = 1 \qquad (2.95)$$

Moreover, all corresponding positive and negative deviations equal 0.2, that is,

$$\Delta\theta_{C_{w2}}^{-} = \Delta\theta_{C_{w2}}^{+} = \theta_{M_{u1}}^{-} = \Delta\theta_{M_{u1}}^{+} = \Delta\theta_{M_{u2}}^{-} = \Delta\theta_{M_{u2}}^{+} = 0.2 \qquad (2.96)$$

By solving the conventional flexibility index model with fixed nominal conditions (i.e., Equation 2.98), the corresponding FI_s can be found to be 0.196. The only active constraint in this solution is associated with the maximum concentration at the inlet of $U2$ (CI_{u2}^{U}).

In this example, the nominal mass loads of both water-using units are assumed to be adjustable and the corresponding multipliers (i.e., $\theta_{M_{u1}}$ and $\theta_{M_{u2}}$) can thus be regarded as type-I parameters in θ_I. Consequently, the only remaining multiplier θ_{Cw2} should be treated as an uncertain parameter in θ_{II}. The DE optimizer has been used to search for the largest possible FI_s of the given water network structure.

Notice that the agent positions in the DE search should be distributed within a region that is bounded by the lower and upper limits of realizable type-I parameters. In the present case, this region is

$$0 \leq \theta_{M_{u1}} \leq \left(\frac{M_{u1}^U}{\overline{M}_{u1}} \right)$$
$$0 \leq \theta_{M_{u2}} \leq \left(\frac{M_{u2}^U}{\overline{M}_{u2}} \right)$$

(2.97)

$FI = \min \delta$

s.t.

$F_{w1} - f_{w1,u1} = 0 (WB_{w1}), F_{w2} - f_{w2,u2} = 0 (WB_{w2}), F_{s1} - f_{t1,s1} = 0 (WB_{s1})$

$F_{u1} - f_{w1,u1} = 0 (WB_{u1}^{in}), F_{u2} - f_{w2,u2} - f_{u1,u2} - f_{t1,u2} = 0 (WB_{u2}^{in}), F_{t1} - f_{u2,t1} = 0 (WB_{t1}^{in})$

$F_{u1} - f_{u1,u2} = 0 (WB_{u1}^{out}), F_{u2} - f_{u2,t1} = 0 (WB_{u2}^{out}), F_{t1} - f_{t1,u2} - f_{t1,s1} = 0 (WB_{t1}^{out})$

$F_{u1}CI_{u1} - f_{w1,u1}C_{w1} = 0 (CB_{u1}), F_{u2}CI_{u2} - f_{w2,u2}\overline{C}w_2\theta_{Cw2} - f_{u1,u2}CO_{u1} - f_{t1,u2}CO_{t1}$

$\qquad = 0 (CB_{u2})$

$F_{t1}CI_{t1} - f_{u2,t1}CO_{u2} = 0 (CB_{t1}), F_{s1}C_{s1} - f_{t1,s1}CO_{t1} = 0 (CB_{s1})$

$F_{u1}(CI_{u1} - CO_{u1}) + \overline{M}_{u1}\theta_{M_{u1}} = 0 (PC_{u1}), F_{u2}(CI_{u2} - CO_{u2}) + \overline{M}_{u2}\theta_{M_{u2}}$

$\qquad = 0 (PC_{u2}), CI_{t1}(1 - RR_{t1}) - CO_{t1} = 0 (PC_{t1})$

$F_{w1} - F_{w1}^U \leq 0, CI_{u1} - CI_{u1}^U \leq 0, CI_{u2} - CI_{u2}^U \leq 0, CO_{u1} - CO_{u1}^U \leq 0$

$CO_{u2} - CO_{u2}^U \leq 0, CI_{t1} - CI_{t1}^U \leq 0, F_{t1} - F_{t1}^U \leq 0, C_{s1} - C_{s1}^U \leq 0$

$\lambda_{F_{w1}^U} - y_{F_{w1}^U} \leq 0, \lambda_{CI_{u1}^U} - y_{CI_{u1}^U} \leq 0, \lambda_{CI_{u2}^U} - y_{CI_{u2}^U} \leq 0, \lambda_{CO_{u1}^U} - y_{CO_{u1}^U} \leq 0$

$\lambda_{CO_{u2}^U} - y_{CO_{u2}^U} \leq 0, \lambda_{CI_{t1}^U} - y_{CI_{t1}^U} \leq 0, \lambda_{F_{t1}^U} - y_{F_{t1}^U} \leq 0, \lambda_{C_{s1}^U} - y_{C_{s1}^U} \leq 0$

$\lambda_{F_{w1}^U} + \lambda_{CI_{u1}^U} + \lambda_{CI_{u2}^U} + \lambda_{CO_{u1}^U} + \lambda_{CO_{u2}^U} + \lambda_{CI_{t1}^U} + \lambda_{F_{t1}^U} + \lambda_{C_{s1}^U} = 1$

$y_{F_{w1}^U} + y_{CI_{u1}^U} + y_{CI_{u2}^U} + y_{CO_{u1}^U} + y_{CO_{u2}^U} + y_{CI_{t1}^U} + y_{F_{t1}^U} + y_{C_{s1}^U} \leq 3$

$SF_{w1}^u - U(1 - y_{F_{w1}^U}) \leq 0, S_{CI_{u1}^U} - U(1 - y_{CI_{u1}^U}) \leq 0, S_{CI_{u2}^U} - U(1 - y_{CI_{u2}^U}) \leq 0, S_{CO_{u1}^U}$

$\qquad - U(1 - y_{CO_{u1}^U}) \leq 0$

$$S_{CO_{u2}^U} - U(1 - y_{CO_{u2}^U}) \le 0, S_{CI_{t1}^U} - U(1 - y_{CI_{t1}^U}) \le 0, S_{F_{t1}^U} - U(1 - y_{F_{t1}^U}) \le 0, S_{C_{s1}^U}$$

$$- U(1 - y_{C_{s1}^U}) \le 0$$

$$\mu_{WB_{w1}} + \lambda_{F_{w1}^U} = 0, -\mu_{WB_{w1}} - \mu_{WB_{u1}^{in}} - C_{w1}\mu_{CB_{u1}} = 0, -\mu_{WB_{w2}} - \mu_{WB_{u2}^{in}} - \bar{C}_{w2}\theta_{C_{w2}}\mu_{CB_{u2}} = 0$$

$$\mu_{WB_{u1}^{in}} + \mu_{WB_{u1}^{out}} + CI_{u1}\mu_{CB_{u1}} + (CI_{u1} - CO_{u1})\mu_{PC_{u1}} = 0$$

$$\mu_{WB_{u2}^{in}} + \mu_{WB_{u2}^{out}} + CI_{u2}\mu_{CB_{u2}} + (CI_{u2} - CO_{u2})\mu_{PC_{u2}} = 0$$

$$\mu_{WB_{t1}^{in}} + \mu_{WB_{t1}^{out}} + CI_{t1}\mu_{CB_{t1}} + \lambda_{F_{t1}^U} = 0, -\mu_{WB_{u2}^{in}} - \mu_{WB_{u2}^{out}} + CO_{u1}\mu_{CB_{u2}} = 0$$

$$-\mu_{WB_{u2}^{in}} - \mu_{WB_{u2}^{out}} + CO_{t1}\mu_{CB_{u2}} + \mu_{ft1,u2} = 0, -\mu_{WB_{t1}^{in}} - \mu_{WB_{u2}^{out}} + CO_{u2}\mu_{CB_{t1}} = 0$$

$$-\mu_{WB_{t1}^{out}} - \mu_{WB_{s1}} + CO_{t1}\mu_{CB_{s1}} = 0, F_{u1}\mu_{CB_{u1}} + F_{u1}\mu_{PC_{u1}} + \lambda_{CI_{u1}^U} = 0$$

$$F_{u2}\mu_{CB_{u2}} + F_{u2}\mu_{PC_{u2}} + \lambda_{CI_{u2}^U} = 0, F_{t1}\mu_{CB_{t1}} + (1 - RR_{t1})\mu_{PC_{t1}} + y_{CI_{t1}^U} = 0$$

$$-f_{u1,u2}\mu_{CB_{u2}} - F_{u1}\mu_{PC_{u1}} + y_{CO_{u1}^U} = 0, -f_{u2,t1}\mu_{CB_{t1}} - F_{u2}\mu_{PC_{u2}} + y_{CO_{u2}^U} = 0$$

$$-f_{t1,s1}\mu_{CB_{s1}} - \mu_{PC_{t1}} = 0, \mu_{WB_{s1}} + C_{s1}\mu_{CB_{s1}} = 0, F_{s1}\mu_{CB_{s1}} + \lambda_{C_{s1}^U} = 0$$

$$0 \le 1 - 0.2\delta \le \theta_{C_{w2}} \le 1 + 0.2\delta$$

$$0 \le \theta_{M_{u1}}^N - 0.2\delta \le \theta_{M_{u1}} \le \theta_{M_{u1}}^N + 0.2\delta \le \left(\frac{M_{u1}^U}{\bar{M}_{u1}} \right)$$

$$0 \le \theta_{M_{u2}}^N - 0.2\delta \le \theta_{M_{u2}} \le \theta_{M_{u2}}^N + 0.2\delta \le \left(\frac{M_{u2}^U}{\bar{M}_{u2}} \right)$$

$$y_{F_{w1}^u}, y_{CI_{u1}^U}, y_{CI_{u2}^U}, y_{CO_{u1}^U}, y_{CO_{u2}^U}, y_{CI_{t1}^U}, y_{F_{t1}^U}, y_{C_{s1}^U} \in \{0,1\}$$

$$S_{F_{w1}^u}, S_{CI_{u1}^U}, S_{CI_{u2}^U}, S_{CO_{u1}^U}, S_{CO_{u2}^U}, S_{CI_{t1}^U}, S_{F_{t1}^U}, S_{C_{s1}^U} \ge 0 \qquad (2.98)$$

For the problem at hand, the maximum allowable number of iteration steps in the search process was set to be 20, and the allowable upper limit of relative error was 10^{-6}. The initial population size was 5, and it took approximately 100 seconds to produce the optimal FI_s (1.6148). According to the optimal solution, it could also be observed that the nominal values of $\theta_{M_{u1}}^N$ and $\theta_{M_{u2}}^N$ should be adjusted to 0.328 and 1.042, respectively. In other words, the nominal mass load of $U1$ needs to be reduced to about 32.8% of the original level (or 0.656 kg/h), whereas that of $U2$ should be 4.2% higher (i.e., 5.21 kg/h). This is clearly reasonable because the active constraint in the optimal solution of the original flexibility index model is associated with the maximum inlet concentration of $U2$.

Riyanto and Chang (2010) suggested that to improve the operational flexibility of a given water network, one can (1) raise the upper limit of the freshwater supply

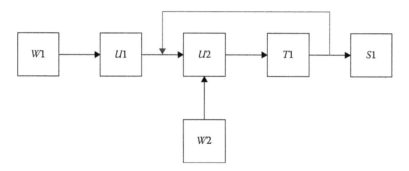

FIGURE 2.7 Revamped water network structure. (Reprinted with permission from Riyanto 2009. Copyright 2009 National Cheng Kung University Library.)

rate and/or (2) modify the network structure. Although successful applications were reported, it should be noted that both approaches inevitably incur extra operating and/or capital costs. Additional case studies are thus presented later to demonstrate the advantages of the current strategy.

Notice first that FI_s can be improved to 0.995 by raising the freshwater supply limit to 45 t/h. However, other than the extra freshwater cost, this improvement is obviously less impressive when compared with the FI_s value achieved by changing the nominal values (1.6148). Next let us consider a revamped structure considered by Riyanto (2009) (see Figure 2.7). Notice that a new pipeline from $T1$ to $U2$ is added to relax the active constraint corresponding to CI_{u2}^U. By solving the conventional flexibility index model with *fixed* nominal conditions, it can be found that such a revamped design improves FI_s to 3.829. In addition, the index value can be further raised to 4.226 by increasing the freshwater supply limit to 45 t / h for this revamped network. It can be observed that the corresponding active constraints are associated with the maximum inlet concentration of U2 (CI_{u2}^U) and the maximum throughput of $T1$ (F_{t1}^U). It should also be noted that the aforementioned improvements can be realized only with additional operating/capital costs. Finally, further enhancement in operational flexibility can be achieved by changing the nominal values of θ_{Mu1}^N and θ_{Mu2}^N to 0.89096 and 0.91773, respectively. In particular, the steady-state flexibility index can be raised to 4.452 without additional investments. These adjustments are obviously quite effective for relaxing the active constraints mentioned earlier.

2.5 NONREPRESENTATIVE FLEXIBILITY MEASURE

Geometrically speaking, the steady-state flexibility index can be regarded as an aggregated measure of the distances between the given nominal point and all faces of the largest inscribable hypercube inside the feasible region. Hence its value may not be a truly representative indicator of the system flexibility when the chosen nominal point is very far off from the center and/or the biggest inscribable hypercube is much smaller than the feasible region due to concavities. For a conceptual understanding, let us consider the simple example studied in Goyal and Ierapetritou (2003):

$$f_1 = 2 - \theta_x \leq 0$$

$$f_2 = -\theta_y - 2 \leq 0$$

$$f_3 = \theta_y - 3 \leq 0$$

$$f_4 = \theta_x - d - \theta_y^2 \leq 0 \tag{2.99}$$

The design variable (d) here is set to 5, whereas θ_x and θ_y are the uncertain parameters. The corresponding feasible region in the parameter space is shown in Figure 2.8. Let us next assume that the permissible hypercube in the parameter space can be defined as follows:

$$\theta_x^N = 4.5, \quad \theta_y^N = 0.0$$

$$\Delta\theta_x^+ = \Delta\theta_x^- = \Delta\theta_y^+ = \Delta\theta_y^- = 4.0 \tag{2.100}$$

To inscribe this hypercube into the feasible region, it is clearly necessary to set $FI_s = \delta = \dfrac{0.5}{4.0} = 0.125$ according to Equation 2.10. Note that the area of the contracted hypercube is only 0.25, which is much smaller than that of the feasible region (26.67).

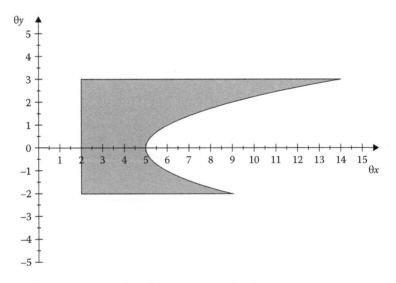

FIGURE 2.8   Nonrepresentative FI_s in a nonconvex feasible region.

REFERENCES

Adi, V.S.K., Chang, C.T., 2011. Two-tier search strategy to identify nominal operating conditions for maximum flexibility. *Industrial & Engineering Chemistry Research* 50, 10707–10716.

Bansal, V., Perkins, J.D., Pistikopoulos, E.N., 2000. Flexibility analysis and design of linear systems by parametric programming. *Aiche Journal* 46, 335–354.

Bansal, V., Perkins, J.D., Pistikopoulos, E.N., 2002. Flexibility analysis and design using a parametric programming framework. *Aiche Journal* 48, 2851–2868.

Biegler, L.T., Grossmann, I.E., Westerberg, A.W., 1997. *Systematic Methods of Chemical Process Design.* Upper Saddle River, NJ: Prentice Hall PTR.

Chang, C.T., Li, B.H., Liou, C.W., 2009. Development of a generalized mixed integer nonlinear programming model for assessing and improving the operational flexibility of water network designs. *Industrial & Engineering Chemistry Research* 48, 3496–3504.

Dimitriadis, V.D., Pistikopoulos, E.N., 1995. Flexibility analysis of dynamic-systems. *Industrial & Engineering Chemistry Research* 34, 4451–4462.

Ferris, M. C, 2005. MATLAB and GAMS: *Interfacing optimization and visualization software.* Computer Sciences Department, University of Wisconsin-Madison, available at http://www.cs.wisc.edu/math-prog/matlab.html.

Floudas, C.A., Gumus, Z.H., Ierapetritou, M.G., 2001. Global optimization in design under uncertainty: Feasibility test and flexibility index problems. *Industrial & Engineering Chemistry Research* 40, 4267–4282.

Goldberg, D.E., 1989. *Genetic Algorithms in Search, Optimization, and Machine Learning.* Reading, MA: Addison-Wesley Publishing Co.

Goyal, V., Ierapetritou, M.G., 2003. Framework for evaluating the feasibility/operability of nonconvex processes. *Aiche Journal* 49, 1233–1240.

Grossmann, I.E., Floudas, C.A., 1987. Active constraint strategy for flexibility analysis in chemical processes. *Computers & Chemical Engineering* 11, 675–693.

Grossmann, I.E., Halemane, K.P., 1982. Decomposition strategy for designing flexible chemical-plants. *Aiche Journal* 28, 686–694.

Halemane, K.P., Grossmann, I.E., 1983. Optimal process design under uncertainty. *Aiche Journal* 29, 425–433.

Holland, J.H., 1975. *Adaptation in Natural and Artificial Systems: An Introductory Analysis with Applications to Biology, Control, and Artificial Intelligence.* Ann Arbor, MI: University of Michigan Press.

Holland, J.H., 1992. *Adaptation in Natural and Artificial Systems: An Introductory Analysis with Applications to Biology, Control, and Artificial Intelligence*, 1st MIT Press ed. Cambridge, MA: MIT Press.

Kuo, Y.C., 2015. Applications of the dynamic and temporal flexibility indices, Department of Chemical Engineering. National Cheng Kung University, Tainan.

Li, B.H., Chang, C.T., 2011. Efficient flexibility assessment procedure for water network designs. *Industrial & Engineering Chemistry Research* 50, 3763–3774.

Lima, F.V., Georgakis, C., 2008. Design of output constraints for model-based non-square controllers using interval operability. *Journal of Process Control* 18, 610–620.

Lima, F.V., Georgakis, C., Smith, J.F., Schnelle, P.D., Vinson, D.R., 2010a. Operability-based determination of feasible control constraints for several high-dimensional nonsquare industrial processes. *Aiche Journal* 56, 1249–1261.

Lima, F.V., Jia, Z., Lerapetritou, M., Georgakis, C., 2010b. Similarities and differences between the concepts of operability and flexibility: The steady-state case. *Aiche Journal* 56, 702–716.

Malcolm, A., Polan, J., Zhang, L., Ogunnaike, B.A., Linninger, A.A., 2007. Integrating systems design and control using dynamic flexibility analysis. *Aiche Journal* 53, 2048–2061.

Ostrovski, G.M., Achenie, L.E.K., Karalapakkam, A.M., Volin, Y.M., 2002. Flexibility analysis of chemical processes: Selected global optimization sub-problems. *Optimization and Engineering* 3, 31–52.

Ostrovsky, G.M., Achenie, L.E.K., Wang, Y.P., Volin, Y.M., 2000. A new algorithm for computing process flexibility. *Industrial & Engineering Chemistry Research* 39, 2368–2377.

Pedersen, M.E.H., 2008. Tuning & simplifying heuristical optimization, School of Engineering Sciences. University of Southampton, Southampton.

Petracci, N.C., Hoch, P.M., Eliceche, A.M., 1996. Flexibility analysis of an ethylene plant. *Computers & Chemical Engineering* 20, S443–S448.

Pistikopoulos, E.N., Grossmann, I.E., 1988a. Evaluation and redesign for improving flexibility in linear-systems with infeasible nominal conditions. *Computers & Chemical Engineering* 12, 841–843.

Pistikopoulos, E.N., Grossmann, I.E., 1988b. Optimal retrofit design for improving process flexibility in linear-systems. *Computers & Chemical Engineering* 12, 719–731.

Pistikopoulos, E.N., Grossmann, I.E., 1989a. Optimal retrofit design for improving process flexibility in nonlinear-systems.1. Fixed degree of flexibility. *Computers & Chemical Engineering* 13, 1003–1016.

Pistikopoulos, E.N., Grossmann, I.E., 1989b. Optimal retrofit design for improving process flexibility in nonlinear-systems.2. Optimal level of flexibility. *Computers & Chemical Engineering* 13, 1087–1096.

Riyanto, E., 2009. A heuristical revamp strategy to improve operational flexibility of existing water networks, Department of Chemical Engineering. National Cheng Kung University, Tainan.

Riyanto, E., Chang, C.T., 2010. A heuristic revamp strategy to improve operational flexibility of water networks based on active constraints. *Chemical Engineering Science* 65, 2758–2770.

Storn, R., Price, K., 1997. Differential evolution - A simple and efficient heuristic for global optimization over continuous spaces. *Journal of Global Optimization* 11, 341–359.

Swaney, R.E., Grossmann, I.E., 1985a. An index for operational flexibility in chemical process design.1. Formulation and theory. *Aiche Journal* 31, 621–630.

Swaney, R.E., Grossmann, I.E., 1985b. An index for operational flexibility in chemical process design.2. Computational algorithms. *Aiche Journal* 31, 631–641.

Varvarezos, D.K., Grossmann, I.E., Biegler, L.T., 1995. A sensitivity based approach for flexibility analysis and design of linear process systems. *Computers & Chemical Engineering* 19, 1301–1316.

Volin, Y.M., Ostrovskii, G.M., 2002. Flexibility analysis of complex technical systems under uncertainty. *Automation and Remote Control* 63, 1123–1136.

3 Volumetric Flexibility Analysis

As indicated in the previous chapter, the steady-state flexibility index (FI_s) can be treated as a suitable performance measure of a given process only when its feasible region is convex. This prerequisite requirement is clearly impractical because a wide variety of chemical engineering models are nonlinear and, thus, nonconvexity is a common feature that cannot be ignored. Geometrically speaking, the aforementioned index FI_s can be regarded as an aggregated indicator of the distances between the given nominal point and all faces of the biggest inscribable hypercube inside the feasible region. Hence, a feasible region may be grossly misrepresented if the chosen nominal point is very far off from the center and/or the biggest inscribable hypercube is much smaller than the feasible region due to concavities. This drawback has been clearly demonstrated in Section 2.5.

Lai and Hui (2008) suggested using an alternative metric—that is, the volumetric flexibility index (denoted in this book as FI_v)—to complement the original approach. Essentially, this metric can be viewed in 3-D as the *volumetric fraction* of the feasible region inside a cube bounded by the expected upper and lower limits of uncertain process parameters. Because the total volume of the feasible region is calculated without needing to specify a nominal point and/or to identify the largest inscribable cube in the feasible region, the magnitude of FI_v can be more closely linked to process flexibility in cases where the feasible regions are nonconvex. However, in practical applications, the feasible regions may be quite complex and sometimes odd shaped. In particular, these geometric objects can be nonconvex, nonsimply connected, and even disconnected in a high-dimensional parameter space (Banerjee and Ierapetritou, 2005; Croft et al., 1994; Krantz, 1999). For any such object, the accuracy of volume estimation depends largely on how well its boundaries can be identified and characterized. In fact, several effective algorithms have already been developed, and a brief summary of their pros and cons are given next.

The *simplicial approximation* approach first proposed by Goyal and Ierapetritou (2003) is not only quite accurate but also capable of handling nonconvex regions; however, its drawbacks can be mainly attributed to the need for *a priori* knowledge of the region shape and repetitive iteration steps to generate the optimal boundary points. The *α-shape surface reconstruction method* (Banerjee and Ierapetritou, 2005) was designed to handle nonconvex and disjoint regions. By implementing this algorithm according to properly sampled points, one can generate a reasonable polygonal representation of the feasible domain. However, the accurate estimate of its hypervolume is attainable only if a suitable α-shape factor can be identified efficiently. Tuning of such an algorithm parameter in realistic applications can be very tricky, especially when the feasible regions are topologically complex; Zilinskas et al. (2006) used sample points that are uniformly distributed over a unit cube to identify the feasible region of a distillation train, but the corresponding hyper-volume could not be quantified easily; Bates et al. (2007) utilized *search cones* to identify the feasible region with uniform sampling points. This approach is especially impractical for odd-shaped regions because it is imperative to strike a proper balance between having enough points to characterize the region well and having too many points, as this can make the model fitting process unstable. By using a fixed number of *auxiliary vectors*, the hypervolume of a feasible region may be quickly determined with accuracy comparable to those of the other methods (Lai and Hui, 2008). Unfortunately, in cases where the nonconvex constraints are present, a serious deterioration in estimation accuracy can occur due to the relatively small number of auxiliary vectors used in computation. On the other hand, the *subspace feasibility test* suggested by the same authors is, in principle, the most accurate numerical strategy for hypervolume estimation if the size of each subspace can be made small enough. However, because these subspaces are created by evenly partitioning the *entire* hypercube bounded between the upper and lower parameter limits, some of the tests do not seem to be necessary if the boundaries of the feasible region can also be taken into consideration. Therefore, as the dimension of parameter space increases, the enormous number of required subspaces can render the computation inefficient.

Finally, notice that it is possible to characterize the feasible region even when the closed-form model of a given process is not available through the use of *surrogate-based feasibility analysis*. The so-called *high-dimensional model representation* (HDMR) has been adopted in Banerjee and Ierapetritou (2002, 2003) and Banerjee et al. (2010) for input–output mapping of such processes, whereas the Kriging-based methodology was later proposed by Boukouvala and Ierapetritou (2012) for essentially the same purpose. Although these methods are based on samples, the developments of surrogate models for black-box problems and problems with known specific models have both been reported in Rogers and Ierapetritou (2015a, b). This strategy involves developing a surrogate model to represent the feasibility function and using it to reproduce the feasible region. Note that the dependency on the sample accuracy and the proper surrogate model are crucial in this practice. Although the surrogate model is often less computationally expensive, it may lack the physical insights needed to identify the potential debottlenecking measures accurately. Notice also that it has always been an attractive incentive for locating the active constraint(s) in traditional flexibility analysis (Grossmann and Floudas, 1987). Moreover, whereas the surrogate models have been applied successfully for feasibility analysis, the resulting index values may not be identical to those of the steady-state flexibility index (FI_s) or the volumetric flexibility index (FI_v) (Rogers and Ierapetritou, 2015a, b). Hence, it is difficult to compare these different approaches on the same basis.

The issues noted earlier in evaluating FI_v can be addressed with an improved computation procedure described in the present chapter. As mentioned previously, the most critical step in this procedure should be concerned with accurate characterization of the feasible region. Specifically, the domain boundaries in parameter space are depicted with proximity points according to a *random line search* algorithm. Two main advantages of this approach are outlined as follows:

- A random search strategy should be inherently more efficient in sketching the operable region without the *a priori* geometric knowledge, for example, see Goyal and Ierapetritou (2003).
- An additional advantage is that two or more boundary points can be produced with a single line search, which makes the proposed strategy much more efficient than the other random search methods (e.g., generating and testing one point at a time).

Note that it is important to obtain sample points near or at boundaries because the interior points are not needed in partitioning a high-dimensional feasible region and then computing its hypervolume. For the former purpose, a Delaunay triangulation technique is applied to create simplexes according to the aforementioned randomly sampled boundary points. The centroid of every simplex is then checked for infeasibility, and the hypervolumes of all the feasible ones can be summed to compute the volumetric flexibility index. This computation strategy can be carried out without repetitively tuning any algorithmic parameter, and the resulting estimates are believed to be more accurate than those obtained with any other existing method with less computation effort.

3.1 FEASIBILITY CHECK

The feasibility check is an essential computation step repeatedly performed at various stages in evaluating FI_v. In order to provide a clear explanation, let us consider the system formulation defined by Equations 2.1 through 2.4 in the previous chapter and

$$\theta^L \leq \theta \leq \theta^U \tag{3.1}$$

where θ^U and θ^L, respectively, represent the vectors of expected upper and lower bounds of the uncertain parameters. Although in principle we can construct a mathematical model for a feasibility check regarding the earlier constraints, it is more convenient to eliminate the state variables from Equation 2.3 then express the inequality constraints in Equation 2.4 as

$$g_j\left(\mathbf{d}, \mathbf{x}(\mathbf{d}, \mathbf{z}, \theta), \mathbf{z}, \theta\right) = f_j\left(\mathbf{d}, \mathbf{z}, \theta\right) \leq 0 \tag{3.2}$$

It should be noted that given a fixed design ($\overline{\mathbf{d}}$) and a particular point (say $\mathbf{b}$) in the parameter hypercube defined by Equation 3.1, the feasibility of this given point cannot be confirmed in a straightforward fashion due to the presence of control variables ($\mathbf{z}$) in the model constraints (i.e., Equations 2.3 and 2.4). A mathematical programming model must be solved for this purpose. Specifically, this optimization problem can be expressed as

$$-P\left(\overline{\mathbf{d}}, \mathbf{b}\right) = \min_{\mathbf{z}} \max_{j \in \mathbb{J}} f_j \tag{3.3}$$

If $P \geq 0$, then the given point in parameter space can be considered feasible, and the corresponding control values will be denoted as $\bar{z}$. Finally, notice that the aforementioned vector $\mathbf{b}$ can be either selected randomly or assigned deterministically, and the upper-level minimization is introduced primarily to eliminate the possibilities of producing erroneous negative P values with the feasible points near boundaries.

3.2   INTEGRATED COMPUTATION STRATEGY

By integrating several software tools, computation of the volumetric flexibility index requires a sequential implementation of five distinct steps: (1) placement of boundary points with a random line search, (2) generation of simplexes using the Delaunay triangulation strategy, (3) removal of infeasible simplexes, (4) calculation of total hypervolume, and (5) evaluation of volumetric flexibility index (FI_v). A brief explanation of each step and an overall flowchart are provided in the sequel.

3.2.1   RANDOM LINE SEARCH

Clearly, a precise geometric characterization of the nonconvex feasible region is the prerequisite of accurate FI_v evaluation. Several alternative strategies have already been developed for this purpose based on the ideas of simplicial approximation (Goyal and Ierapetritou, 2003), α-shape surface reconstruction (Banerjee and Ierapetritou, 2005), and auxiliary vector and subspace feasibility test (Lai and Hui, 2008). Because, as mentioned before, these available methods are still not satisfactory for practical applications, a random line search algorithm is presented here for placing feasible points at the boundaries of the feasible region in the parameter space. Specifically, this search is realized by solving a mathematical programming model described later.

Let us first assume that a feasibility check has already been performed for a given design $\bar{\mathbf{d}}$ on a randomly generated vector $\mathbf{b}$ to obtain $\bar{\mathbf{z}}$. A subsequent mathematical program can then be formulated accordingly after producing still another random vector $\mathbf{a}$, that is,

$$Q\left(\bar{\mathbf{d}}, \bar{\mathbf{z}}, \mathbf{a}, \mathbf{b}\right) = \min_{t, j \in \mathbb{J}} s_j$$

$s.t.$

$$f_j\left(\bar{\mathbf{d}}, \bar{\mathbf{z}}, \theta\right) + s_j = 0, \quad s_j \geq 0, \quad j \in \mathbb{J}$$

$$\theta = \mathbf{a}t + \mathbf{b}, \quad +\infty > t > -\infty$$

(3.4)

where s_j is the slack variable of inequality $j \in \mathbb{J}$ and t is the parametric variable for a straight line. If the optimal objective value is smaller than a designated threshold value ε, that is, $Q < \varepsilon$, then the corresponding vector(s) in the parameter space should be regarded as proximity point(s) of the boundaries. To enhance search efficiency, it may also be beneficial to make multiple line searches, that is, generate more than one random vector **a**, on the basis of the same feasible point **b**. The maximum number of lines allowed per feasible point is denoted as N_{line}, whereas the total number of proximity points targeted in the search is denoted as N_{point}. The corresponding search procedure can be outlined as follows:

1. Let $n = 0$ and $\Theta = \varnothing$.
2. Generate a random vector **b** within the parameter hypercube according to Equation 3.1. Perform a feasibility check on **b** by solving Equation 3.3.
3. If $P < 0$, then go to step 2. Otherwise, save **b** and $\bar{z}$ and then go the next step.
4. Let $k = 0$ and $\Omega = \varnothing$.
5. Generate an additional random vector **a** and solve Equation 3.4.
6. If $Q \geq \varepsilon$, then go to step 5. Otherwise, incorporate all solutions (i.e., the proximity points) into the set Ω and go to the next step.
7. Let $k = k + 1$. If $k \leq N_{line}$, go to step 5. Otherwise, go to the next step.
8. Let $\Theta = \Theta \cup \Omega$ and $n = n + \mathrm{card}(\Omega)$, where $\mathrm{card}(\Omega)$ denotes the cardinality of set Ω. If $n \leq N_{point}$, go to step 2. Otherwise, stop.

An obvious advantage of this search strategy is that at least two proximity points can be generated with a single line. For a conceptual understanding, let us consider a motivating example with the following five inequality constraints:

$$f_1 = \theta_2 - 2\theta_1 - 15 \leq 0$$

$$f_2 = \frac{\theta_1^2}{2} + 4\theta_1 - 5 - \theta_2 \leq 0$$

$$f_3 = 10 - \frac{(\theta_1 - 4)^2}{5} - \frac{\theta_2}{0.5} \leq 0 \qquad (3.5)$$

$$f_4 = \theta_2 - 15 \leq 0$$

$$f_5 = \theta_2(6 + \theta_1) - 80 \leq 0$$

Figure 3.1 shows the corresponding feasible region, and note that there are two nonconvex constraints: f_3 and f_5. In the original problem statement (Goyal and Ierapetritou, 2003), a nominal point of $(\theta_1^N, \theta_2^N) = (-2.5, 0)$ was adopted with expected deviations of $(\Delta\theta_1^+, \Delta\theta_1^-) = (7.5, 7.5)$ and $(\Delta\theta_2^+, \Delta\theta_2^-) = (15, 15)$. In this figure, a single random line is drawn to show the possibility of getting multiple solutions with the same **a** and **b**. The proposed search algorithm was coded and implemented on the MATLAB™-GAMS platform (Dirkse et al., 2014; MathWorks, 2016e; Rosenthal, 2016) to produce 1000 boundary points with $N_{point} = 1000$ and $N_{line} = 1$ (see Figure 3.2). It can be observed that the boundaries of the feasible region are well characterized, and, thus an accurate estimate of its boundary may be obtained accordingly.

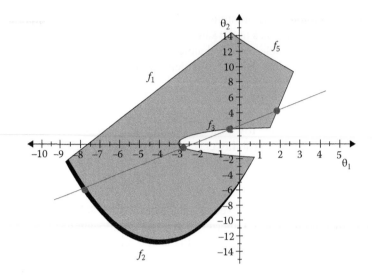

FIGURE 3.1 Feasible region of the motivating example (Reprinted from *Chemical Engineering Science*, 147, Adi, V.S.K. et al., An effective computation strategy for assessing operational flexibility of high-dimensional systems with complicated feasible regions, 137–149. Copyright 2016, with permission from Elsevier Ltd.)

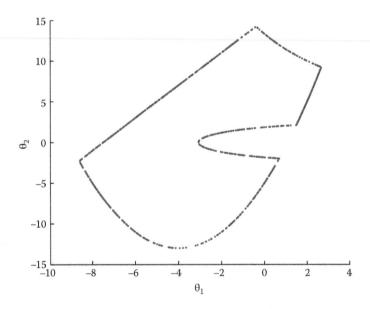

FIGURE 3.2 Proximity points generated in the motivating example. (Reprinted from *Chemical Engineering Science*, 147, Adi, V.S.K. et al., An effective computation strategy for assessing operational flexibility of high-dimensional systems with complicated feasible regions, 137–149. Copyright 2016, with permission from Elsevier Ltd.)

3.2.2 Delaunay Triangulation

The Delaunay triangulation strategy has been widely adopted for scientific computing in diverse applications. Although there are many other computer algorithms available, it is its favorable geometric properties that make this particular one useful for the present purpose. The constrained Delaunay triangulation strategy can, in fact, be implemented in high dimensions without any difficulties, as the algorithm is quite mature and has already been embedded in commercial software (Barber et al., 1996; MathWorks, 2016c). The two-dimensional data set presented previously in Figure 3.2 is again used here as an example for illustration. The simplexes (triangles) shown in Figure 3.3 were obtained by direct implementation of the Delaunay procedure using the MATLAB built-in function "delaunayn" (MathWorks, 2016d). Note that some of the simplexes are located outside the feasible region.

3.2.3 Infeasible Simplexes

For illustration convenience, let us assume that coordinate data of the aforementioned proximity points in n-dimensional space can be stored in a N_{point}-by-n matrix (which is referred to as $\mathbf{X}$), and each row vector of this matrix represents one such point. The

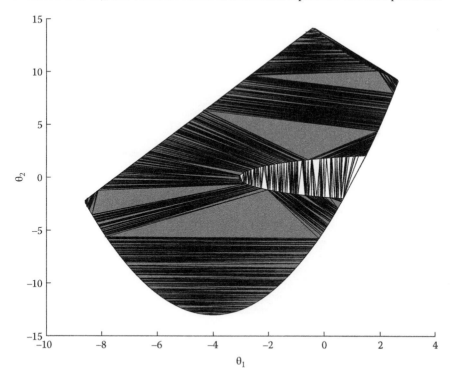

FIGURE 3.3 Delaunay triangulation scheme obtained from boundary points in Figure 3.2. (Reprinted from *Chemical Engineering Science*, 147, Adi, V.S.K. et al., An effective computation strategy for assessing operational flexibility of high-dimensional systems with complicated feasible regions, 137–149, Copyright 2016, with permission from Elsevier Ltd.)

MATLAB function "delaunayn" basically generates a list of N_{simp} simplexes in such a way that no data points in $\mathbf{X}$ are located inside the circumsphere of any simplex (Delaunayn, 2014). This Delaunay triangulation list T is essentially a N_{simp}-by-$n+1$ array in which each row contains the row indices of $\mathbf{X}$ for the $n + 1$ vertices of a simplex; that is, simplex k in the triangulation scheme is uniquely associated with the kth row of list T and its ith element T_i^k $(i = 0, 1, \cdots, n)$ is with row vector T_i^k in $\mathbf{X}$.

Consequently, the centroid $\overline{\theta}^k$ of simplex k can be calculated using the simple formula (Johnson, 2007) given as

$$\overline{\theta}^k = \frac{1}{n+1} \sum_{i=0}^{n} \mathbf{x}_i^k \tag{3.6}$$

where $\mathbf{x}_i^k$ denotes row vector T_i^k in $\mathbf{X}$, that is, the coordinate vector for vertex i of simplex k. Given a fixed design $(\overline{\mathbf{d}})$ and a centroid $\overline{\theta}^k$, the mathematical programming model given in Equation 3.3 can be solved to determine the feasibility of the simplex k with the corresponding centroid $\overline{\theta}^k$. A simplex can be retained only when the corresponding $P \geq 0$. Figure 3.4 shows the enhanced triangulation scheme obtained by removing the infeasible simplexes in Figure 3.3.

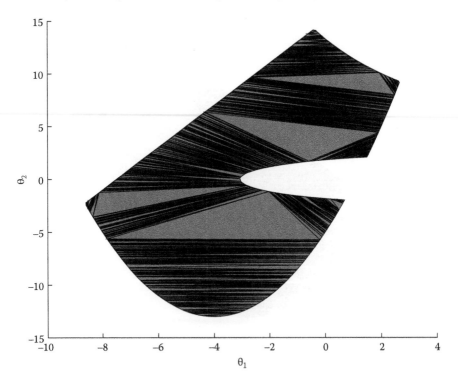

FIGURE 3.4 Enhanced triangulation scheme obtained from Figure 3.3. (Reprinted from *Chemical Engineering Science*, 147, Adi, V.S.K. et al., An effective computation strategy for assessing operational flexibility of high-dimensional systems with complicated feasible regions, 137–149. Copyright 2016, with permission from Elsevier. Ltd.)

3.2.4 HYPERVOLUME CALCULATION

The hypervolume of simplex $k(k = 1,2,\cdots,N_{simp})$ can be computed simply with the following formula (Burkardt, 2013; Stein, 1966):

$$V^k = \frac{1}{n!}\left|\det\left[\ \left(\mathbf{x}_1^k - \mathbf{x}_0^k\right)^T \quad \left(\mathbf{x}_2^k - \mathbf{x}_0^k\right)^T \quad \cdots \quad \left(\mathbf{x}_n^k - \mathbf{x}_0^k\right)^T\ \right]\right| \tag{3.7}$$

where each column of the n-by-n matrix is the transpose of difference between two row vectors representing vertex i ($i = 1,2,\cdots,n$) and the reference vertex $\mathbf{x}_0^k$, respectively. The total hypervolume of the feasible region V_{fr} can therefore be calculated easily by summing those of all feasible simplexes. For the motivating example, analytical integration can be performed and the theoretical area of the feasible region can be determined to be 152.76 units. On the other hand, the total area of all triangles in Figure 3.3 can be found to be 165.93 units by making use of Equation 3.7, whereas that of the feasible region in Figure 3.4 is 152.7442 units (which is 99.99% of the theoretical value).

3.2.5 FLEXIBILITY MEASURE

According to Lai and Hui (2008), the volumetric flexibility index FI_v should be calculated according to the formula:

$$FI_v = \frac{V_{fr}}{V_{ur}} \tag{3.8}$$

where V_{ur} is the hypercube volume bounded by the expected upper and lower limits of uncertain process parameters. Thus, the exact value of volumetric flexibility index for the motivating example is

$$FI_v = \frac{152.76}{(7.5+7.5)\times(15+15)} = 0.34 \tag{3.9}$$

Because this problem has already been solved in several previous studies, it is therefore necessary to first present a summary of all available results:

- The approximated area of the feasible region was found by Goyal and Ierapetritou (2003) with the simplicial approximation approach to be 129.69 units (which is only 84.90% of the theoretical value) and therefore the corresponding estimate of FI_v is 0.29.
- Using the α-shape surface reconstruction algorithm, Banerjee and Ierapetritou (2005) only reported the sampled feasible points without the resulting area. For comparison purposes, the α-shape surface reconstruction computation has been repeated in this study with the built-in MATLAB function "alphaShape" (MathWorks, 2016a) by 3356 evenly distributed points and a critical alpha radius of 0.1694 (MathWorks, 2016b). The estimated area in this case is 148.58 units (which is 97.26% of the theoretical value), and thus, the corresponding FI_v is 0.33.

- Lai and Hui (2008) also studied the same problem by using the auxiliary vector approach with two alternative objectives: (1) maximizing the sum of lengths of the position vectors that represent the interception points and (2) maximizing the sum of squares of the distances between interception points and a reference point. The former yielded an area estimate of 148.03 units (96.90% of the theoretical value) and the corresponding FI_v is 0.33, whereas the latter produced an overestimated area of 155.75 units (101.96% of the theoretical value) and an optimistic flexibility index of 0.35.

As mentioned before, the total area of feasible simplexes in Figure 3.4 was found to be 152.7442 (99.99% of the theoretical value) and, thus, the resulting flexibility measure FI_v (0.34) should be more accurate than any of the aforementioned methods.

3.2.6 OVERALL COMPUTATION FLOWCHART

The earlier algorithms can be integrated into a single flowchart for evaluating the volumetric flexibility index (see Figure 3.5). The optimization runs performed by GAMS are marked with blocks enclosed by dark grey rectangles in this figure, and the computations carried out with MATLAB codes are placed in blocks against the white background. Although this flowchart is self-explanatory, its steps are still briefly described as follows for the sake of illustration completeness: (1) The first step is the random line search. On the MATLAB-GAMS platform, the corresponding computations produce the data set Θ that contains all required feasible proximity points. (2) The feasible proximity points in Θ are then triangulated using the MATLAB N-D Delaunay triangulation built-in function "delaunayn" to generate the triangulation list T. (3) The centroid of every simplex in T is checked for feasibility and the infeasible ones are deleted. The triangulation list T is then updated accordingly. (4) The hypervolumes of all simplexes in the updated list T are computed and summed to estimate the hypervolume of the feasible region. (5) The corresponding volumetric flexibility index is finally evaluated according to its definition.

3.3 DEMONSTRATIVE EXAMPLES

The following examples were selected to demonstrate the effectiveness of the proposed computation strategy in handling disjoint, nonsimply connected, and high-dimensional feasible regions. The flowchart in Figure 3.5 was implemented on a computer system with the following specifications: Acer Veriton P530 F2, 2x Intel® Xeon® CPU E5-2620 v2@2.10GHz (12 cores), 64 GB RAM, Windows 10 64bit, MATLAB 2015b, and GAMS 24.5.3 (August 2015). The default values of ε and N_{line} for the line search were chosen to be 10^{-6} and 1, respectively. In all cases reported later, the time needed for each MATLAB-GAMS call was less than 0.001 second. For the 7D problem in Section 3.3.6, the elapsed time for triangulation was less than 2 hours, and the entire computation process lasted less than 5 hours.

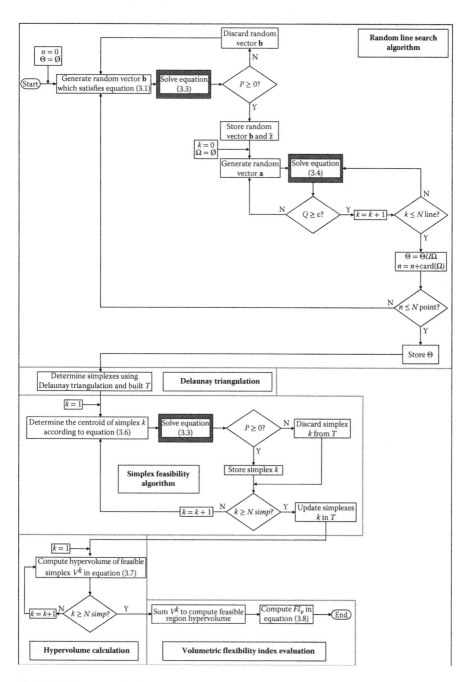

FIGURE 3.5 Overall flowchart. (Reprinted from *Chemical Engineering Science*, 147, Adi, V.S.K. et al., An effective computation strategy for assessing operational flexibility of high-dimensional systems with complicated feasible regions, 137–149. Copyright 2016, with permission from Elsevier, Ltd.)

3.3.1 DISJOINT REGION

To demonstrate the capability of the proposed approach in evaluating disconnected feasible regions, let us consider the heat exchanger network (HEN) in Figure 3.6 (Grossmann and Floudas, 1987). The model formulation was obtained by eliminating the state variables with the equality constraints:

$$f_1 = -25F_{H1} + Q_c - 0.5Q_c F_{H1} + 10 \leq 0$$

$$f_2 = -190F_{H1} + Q_c + 10 \leq 0$$

$$f_3 = -270F_{H1} + Q_c + 250 \leq 0 \qquad\qquad (3.10)$$

$$f_4 = 260F_{H1} - Q_c - 250$$

Q_c in these constraints is the cooling load, which has been treated as a positive-valued *control variable*, whereas F_{H1} is the heat capacity flow rate of hot stream H1, and it is regarded as the only *uncertain parameter* in this problem. It is also assumed that the uncertain parameter has a nominal value of 1.4 kW/K (F_{H1}^N) and the expected positive and negative deviations (i.e., ΔF_{H1}^+ and ΔF_{H1}^-) are both set at 0.4 kW/K.

In the space formed by both the control variable and the uncertain parameter (see Figure 3.7), the *expanded* feasible region defined by Equation 3.10 consists of two disconnected domains (Grossmann and Floudas, 1987). Note that its specific total area, that is, 3.15 units, can be determined by analytical integration. Note also that the simplicial approximation approach requires *a priori* identification of the nonconvex constraint(s), that is, f_1, that causes the division of the feasible region. The total

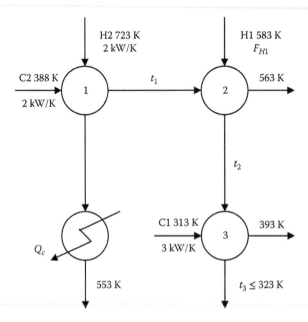

FIGURE 3.6   Heat exchanger network studied in Grossmann and Floudas (1987).

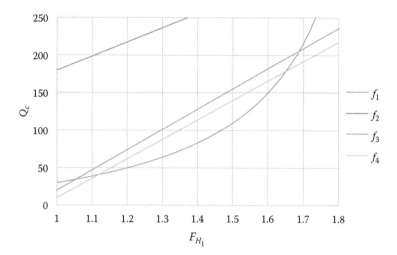

FIGURE 3.7 The expanded feasible region defined by Equation 3.10.

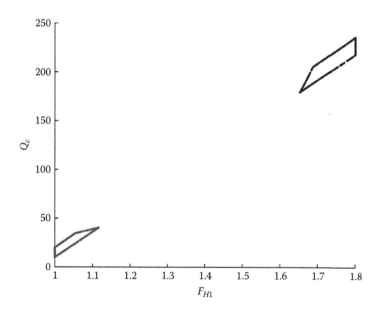

FIGURE 3.8 Boundary points generated for the expanded feasible region. (Reprinted from *Chemical Engineering Science*, 147, Adi, V.S.K. et al., An effective computation strategy for assessing operational flexibility of high-dimensional systems with complicated feasible regions, 137–149. Copyright 2016, with permission from Elsevier, Ltd.)

area of disjoint feasible regions was estimated by Goyal and Ierapetritou (2003) to be 1.84 units, which is 58.41% of the actual value. On the other hand, the proposed search algorithm has also been implemented to characterize the expanded feasible region mentioned earlier. A total of 1000 boundary points were generated by treating Q_c as a *pseudo-parameter* to mimic the feasible boundaries (see Figure 3.8).

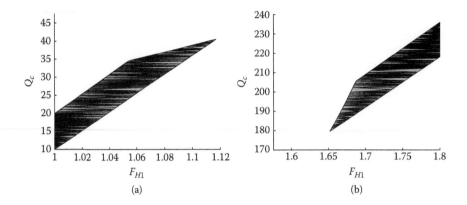

FIGURE 3.9 Delaunay triangulation schemes obtained from Figure 3.7: (a) left; (b) right. (Reprinted from *Chemical Engineering Science*, 147, Adi, V.S.K. et al., An effective computation strategy for assessing operational flexibility of high-dimensional systems with complicated feasible regions, 137–149. Copyright 2016, with permission from Elsevier, Ltd.).

The simplexes shown in Figure 3.9 were obtained by Delaunay triangulation strategy, and the corresponding total area is 3.14 units (99.68% of the actual value).

Finally, it should be noted that the present problem is 1-D because there is only one uncertain parameter F_{H1}. The actual feasible region is just two separated line segments, which can be generated by projecting the expanded region onto the F_{H1} axis. Specifically, from the intersection points of f_1 and f_4, one can easily locate the upper limit of the segment on the left and the lower limit on the right, and their values are 1.118 and 1.651, respectively. Therefore, the exact value of 1-D FI_v is

$$FI_v = \frac{(1.118-1)+(1.8-1.651)}{1.8} = 0.148 \tag{3.11}$$

Based on the data points generated from a random search (see Figure 3.9), the first part spans the interval [1, 1.117] and the second [1.652, 1.8]. Thus the corresponding value of FI_v is

$$FI_v = \frac{(1.117-1)+(1.8-1.652)}{1.8} = 0.147 \tag{3.12}$$

which corresponds to 99.32% of the actual FI_v. This result shows that the proposed search algorithm produces a reliable high-accuracy prediction of the feasible region.

3.3.2 Nonsimply Connected 2-D Region

To show the effectiveness of the proposed approach in handling the nonsimply connected regions, let us revisit the motivating example and introduce an additional constraint:

$$f_6 = -(\theta_1+5)^2 - (\theta_2+5)^2 + 2 \leq 0 \tag{3.13}$$

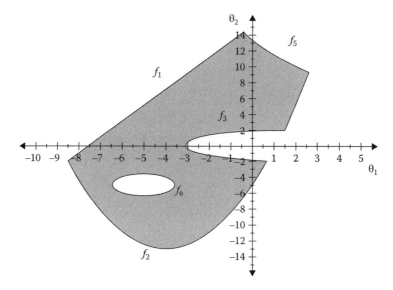

FIGURE 3.10 Feasible region of the 2-D nonsimply connected example. (Reprinted from *Chemical Engineering Science*, 147, Adi, V.S.K. et al., An effective computation strategy for assessing operational flexibility of high-dimensional systems with complicated feasible regions, 137–149. Copyright 2016, with permission from Elsevier, Ltd.)

Figure 3.10 shows the corresponding feasible region. By performing analytical integration, the exact area of this region can be determined to be 146.48 units. Thus, the actual value of volumetric flexibility index should be

$$FI_v = \frac{146.48}{(7.5 + 7.5) \times (15 + 15)} = 0.3255 \tag{3.14}$$

By using the proposed search algorithm, the feasible region was characterized by 1000 boundary points (see Figure 3.11), and the corresponding triangulation scheme is given in Figure 3.12. The area of the feasible region was found to be 146.46 (99.99% of the exact value). Thus the value of FI_v is

$$FI_v = \frac{146.46}{(7.5 + 7.5) \times (15 + 15)} = 0.3255 \tag{3.15}$$

From these results, it can be observed that the proposed algorithms can be easily implemented to produce accurate area estimates of nonsimply connected regions— at least in two-dimensional problems. On the other hand, although the simplicial approximation approach proposed by Goyal and Ierapetritou (2003) is also capable of handling such regions, it is still necessary to identify the special geometric features of f_3, f_5, and f_6 in advance and to construct the simplicial convex hull and outer convex polytope (Goyal and Ierapetritou, 2003).

For comparison purposes, the α-shape surface reconstruction computation (Banerjee and Ierapetritou, 2005) has also been repeated by 3217 evenly distributed points. The feasible area in this case is underestimated to be 141.23 units

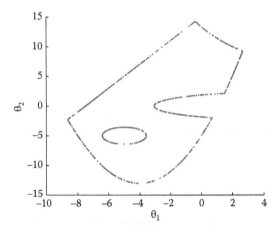

FIGURE 3.11 10^3 boundary points generated in the 2-D nonsimply connected example. (Reprinted from *Chemical Engineering Science*, 147, Adi, V.S.K. et al., An effective computation strategy for assessing operational flexibility of high-dimensional systems with complicated feasible regions, 137–149. Copyright 2016, with permission from Elsevier, Ltd.).

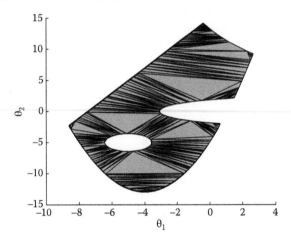

FIGURE 3.12 Delaunay triangulation scheme obtained from Figure 3.10. (Reprinted from *Chemical Engineering Science*, 147, Adi, V.S.K. et al., An effective computation strategy for assessing operational flexibility of high-dimensional systems with complicated feasible regions, 137–149. Copyright 2016, with permission from Elsevier, Ltd.)

(96.42% of the theoretical value), and thus, the corresponding FI_v is 0.31. Although the α-shape surface reconstruction method was designed to handle the nonconvex and nonsimply connected region, the estimation of its hypervolume is attainable only if a suitable α-shape factor can be identified properly. Furthermore, with more sampling points used (3217 points vs. 1000 points), the estimation accuracy is actually lower than that achieved with the proposed method (96.42% vs. 99.99% of the exact value).

3.3.3 Nonconvex 3-D Regions

To show the effects of increasing boundary points, let us next consider a three-parameter feasible region bounded between a cube:

$$f_1 = \theta_x - 3 \leq 0$$

$$f_2 = \theta_y - 3 \leq 0$$

$$f_3 = \theta_z - 3 \leq 0$$

$$f_4 = -\theta_x \leq 0 \qquad (3.16)$$

$$f_5 = -\theta_y \leq 0$$

$$f_6 = -\theta_z \leq 0$$

and also a sphere:

$$f_7 = 1 - (\theta_x - 1.5)^2 - (\theta_y - 1.5)^2 - (\theta_z - 1.5)^2 \leq 0 \qquad (3.17)$$

where θ_x, θ_y, and θ_z are the uncertain parameters considered in the present example. The nominal point was placed at $\theta_x^N = \theta_y^N = \theta_z^N = 1.5$, and the expected positive and negative deviations in these uncertain parameters were chosen to be

$$\Delta\theta_x^+ = \Delta\theta_y^+ = \Delta\theta_z^+ = 2.5$$

$$\Delta\theta_x^- = \Delta\theta_y^- = \Delta\theta_z^- = 1.5 \qquad (3.18)$$

The volume of this feasible region can be determined analytically to be 22.81 units, and thus the exact value of FI_v is 0.36.

Using the auxiliary vector approach (Lai and Hui, 2008), the spherical void inside the cube cannot be detected properly, and thus, an erroneous volume of 27 units was obtained. Note that the auxiliary vector approach calls for two objectives: (1) maximizing the sum of lengths of the position vectors that represent the interception points and (2) maximizing the sum of squares of the distances between interception points and a reference point. The maximum length of the position vectors will be the corners of the cube. Thus, the sphere void is inevitably left undetected. If the other available methods, that is, simplicial approximation and α-shape surface reconstruction, are applied in this case, the drawbacks described in the previous 2-D example can also be observed. The former calls for *a priori* knowledge of the geometric features of the feasible region and tedious steps to construct the simplicial convex hull and outer convex polytope, whereas the latter requires iterative tuning of the algorithm parameter but often yields lower accuracy.

By carrying out the proposed computation procedure, the feasible region can be described accurately. To show the resolution improvement achieved by increasing boundary points, 10^2, 10^3, and 10^4 randomly generated points and their corresponding triangulation schemes are plotted in Figures 3.13 through 3.15, respectively. The total volume of all simplexes in the most refined scheme in Figure 3.15 has been calculated and used as the estimated volume of the feasible region. This estimate for 10^4 boundary points is 22.83 units, which is 100.09% of the theoretical value, and the corresponding FI_v (0.36) is essentially the same as that determined analytically.

With fewer boundary points, that is, 10^2 and 10^3 points, the volume of the feasible region was estimated to be 18.53 units (81.24% of the theoretical value) and 22.92 (100.39% of the theoretical value), respectively. Note that when compared with the results generated with the auxiliary vector approach (Lai and Hui, 2008), the boundaries of the feasible region can be better characterized and the corresponding

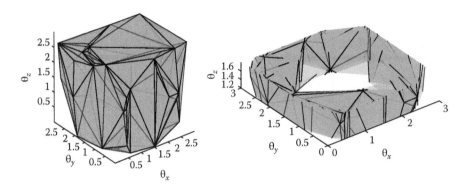

FIGURE 3.13 Left: Triangulation scheme constructed according to 10^2 randomly generated boundary points. Right: Partial triangulation scheme is shown for a limited range of θ_z. (Reprinted from *Chemical Engineering Science*, 147, Adi, V.S.K. et al., An effective computation strategy for assessing operational flexibility of high-dimensional systems with complicated feasible regions, 137–149. Copyright 2016, with permission from Elsevier, Ltd.)

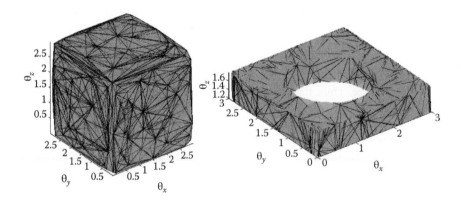

FIGURE 3.14 Left: Triangulation scheme constructed according to 10^3 randomly generated boundary points. Right: Partial triangulation scheme is shown for a limited range of θ_z. (Reprinted from *Chemical Engineering Science*, 147, Adi, V.S.K. et al., An effective computation strategy for assessing operational flexibility of high-dimensional systems with complicated feasible regions, 137–149. Copyright 2016, with permission from Elsevier, Ltd.)

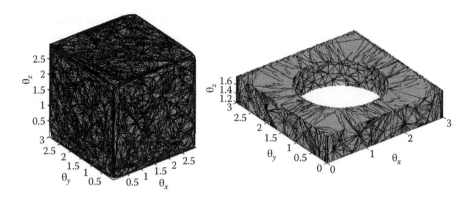

FIGURE 3.15 Left: Triangulation scheme constructed according to 10^4 randomly generated boundary points. Right: Partial triangulation scheme is shown for a limited range of θ_z. (Reprinted from *Chemical Engineering Science*, 147, Adi, V.S.K. et al., An effective computation strategy for assessing operational flexibility of high-dimensional systems with complicated feasible regions, 137–149. Copyright 2016, with permission from Elsevier, Ltd.)

volume more accurately estimated by following the proposed computation procedure with 10^3 points (27.0 vs. 22.8 units). On the other hand, because it is obvious that 10^2 boundary points cannot cover the entire feasible region adequately, as shown in Figure 3.13, the corresponding results are not satisfactory.

Finally, it can be observed from the results obtained in this 3-D problem and other extensive case studies that the feasible region can usually be better characterized with more boundary points until reaching a saturation level. A heuristic rule can thus be deduced to facilitate proper selection of the number of boundary points, that is, this number should at least be set at 10^n (where n is the dimension of parameter space) for rough estimations and may be raised to 10^{n+1} if a higher accuracy is desired. This suggested rule will also be tested in the subsequent examples.

For a direct comparison with the previous work, let us next consider the following three-parameter system studied by Goyal and Ierapetritou (2003):

$$f_1 = \theta_x - 3 \leq 0$$

$$f_2 = \theta_y - 3 \leq 0$$

$$f_3 = \theta_z - 3 \leq 0$$

$$f_4 = 1 - \theta_x^2 - \theta_y^2 - \theta_z^2 \leq 0 \tag{3.19}$$

$$f_5 = -\theta_x \leq 0$$

$$f_6 = -\theta_y \leq 0$$

$$f_7 = -\theta_z \leq 0$$

where the center of the sphere the in previous example is located at the origin, that is, (0, 0, 0). The nominal point was placed similarly at $\theta_x^N = \theta_y^N = \theta_z^N = 1.5$, and the expected positive and negative deviations in the uncertain parameters were chosen to be

$$\Delta\theta_x^+ = \Delta\theta_y^+ = \Delta\theta_z^+ = 2.5$$

$$\Delta\theta_x^- = \Delta\theta_y^- = \Delta\theta_z^- = 1.5$$

(3.20)

The volume of this region can be determined analytically to be 26.48 units, and thus the exact value of FI_v is 0.41.

The volume of the feasible region was originally estimated to be 25.88 units (97.73% of the theoretical value) by Goyal and Ierapetritou (2003) and thus a conservative FI_v value of 0.40 was obtained. On the other hand, Lai and Hui (2008) studied the same problem and produced an even smaller volume estimate, that is, 25.26 units (95.39% of the theoretical value) with the auxiliary vector approach, and a more conservative flexibility index (0.39).

By implementing the proposed method, the feasible region can be characterized by the 10^4 boundary points, and the corresponding triangulation schemes are plotted in Figure 3.16. The total volume of all simplexes has been calculated and used as the estimated volume of the feasible region. This estimate for 10^4 boundary points is 26.47 units, which is 99.96% of the theoretical value, and the corresponding FI_v (0.41) is essentially the same as that determined analytically. Note that when compared with

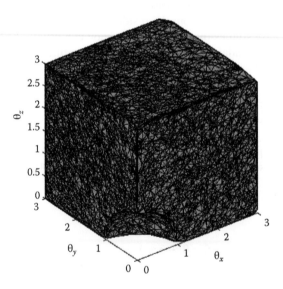

FIGURE 3.16 Triangulation scheme constructed according to 10^4 randomly generated boundary points. (Reprinted from *Chemical Engineering Science*, 147, Adi, V.S.K. et al., An effective computation strategy for assessing operational flexibility of high-dimensional systems with complicated feasible regions, 137–149. Copyright 2016, with permission from Elsevier, Ltd.)

the results reported in Goyal and Ierapetritou (2003) and Lai and Hui (2008), the boundaries of the feasible region can be better characterized and the corresponding volume more accurately estimated by following the proposed computation procedure.

3.3.4 COMPLICATED 3-D REGION

Let us consider another fictitious case when the feasible region is defined in a very complex way as follows (Klaus, 2010):

$$\left\{-8\left(\theta_x^2+\theta_y^2\right)^2\left(\theta_x^2+\theta_y^2+1+\theta_z^2+a^2-b^2\right)+4a^2\left[2\left(\theta_x^2+\theta_y^2\right)^2\right.\right.$$

$$\left.-\left(\theta_x^3-3\theta_x\theta_y^2\right)\left(\theta_x^2+\theta_y^2+1\right)\right]+8a^2\left(3\theta_x^2\theta_y-\theta_y^3\right)\theta_z+4a^2\left(\theta_x^3-3\theta_x\theta_y^2\right)\theta_z^2\right\}^2$$

$$-\left(\theta_x^2+\theta_y^2\right)\left\{2\left(\theta_x^2+\theta_y^2\right)\left(\theta_x^2+\theta_y^2+1+\theta_z^2+a^2-b^2\right)^2+8\left(\theta_x^2+\theta_y^2\right)^2\right.$$

$$+4a^2\left[2\left(\theta_x^3-3\theta_x\theta_y^2\right)-\left(\theta_x^2+\theta_y^2\right)\left(\theta_x^2+\theta_y^2+1\right)\right]-8a^2\left(3\theta_x^2\theta_y-\theta_y^3\right)\theta_z$$

$$-4\left(\theta_x^2+\theta_y^2\right)a^2\theta_z^2\right\}^2=0 \tag{3.21}$$

$$a=2.5$$

$$0\le b\le 0.2$$

$$-1.5\le\theta_x\le 1.5$$

$$-1.5\le\theta_y\le 1.5$$

$$-0.5\le\theta_z\le 0.5$$

This feasible region is an object called a *trefoil*. It will be very difficult to construct the simplicial convex hull and outer convex polytope (Goyal and Ierapetritou, 2003) based on Equation 3.21. It will also be erroneous with the auxiliary vector approach (Lai and Hui, 2008) because the method is limited by the number of search vectors available.

By using the proposed method, the feasible region can be characterized with the 10^4 boundary points, and the corresponding triangulation schemes are plotted in Figures 3.17 and 3.18, respectively. The area of the feasible region was found to be 1.54, and the value of FI_v is

$$FI_v=\frac{1.54}{(1.5+1.5)\times(1.5+1.5)\times(0.5+0.5)}=0.1711 \tag{3.22}$$

Finally, the α-shape surface reconstruction operation (Banerjee and Ierapetritou, 2005) has also been repeated by the 10^4 boundary points generated by the proposed search algorithm. As shown in Figure 3.19, this approach only yielded a poorly characterized feasible region.

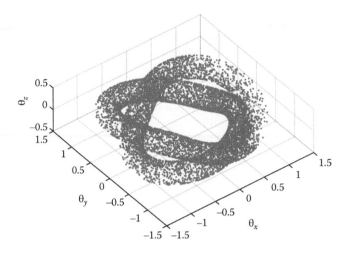

FIGURE 3.17 10^4 boundary points generated in the 3-D complicated feasible region. (Reprinted from *Chemical Engineering Science*, 147, Adi, V.S.K. et al., An effective computation strategy for assessing operational flexibility of high-dimensional systems with complicated feasible regions, 137–149. Copyright 2016, with permission from Elsevier, Ltd.).

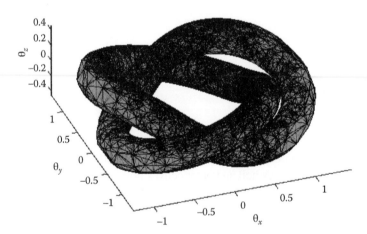

FIGURE 3.18   Triangulation scheme constructed according to 10^4 randomly generated boundary points. (Reprinted from *Chemical Engineering Science*, 147, Adi, V.S.K. et al., An effective computation strategy for assessing operational flexibility of high-dimensional systems with complicated feasible regions, 137–149. Copyright 2016, with permission from Elsevier, Ltd.)

3.3.5   HEAT EXCHANGER NETWORK WITH MULTIPLE UNCERTAIN PARAMETERS

To further demonstrate the effects of increasing boundary points and the validity of the suggested heuristic rule, let us consider the HEN design problem reported in

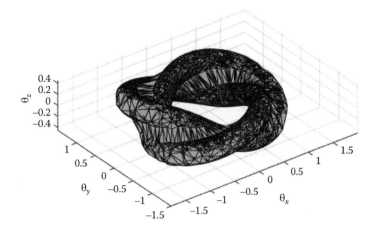

FIGURE 3.19   The feasible region obtained with α-shape surface reconstruction. (Reprinted from *Chemical Engineering Science*, 147, Adi, V.S.K. et al., An effective computation strategy for assessing operational flexibility of high-dimensional systems with complicated feasible regions, 137–149. Copyright 2016, with permission from Elsevier, Ltd.)

Grossmann and Floudas (1987). The inequality constraints imposed in the original model are summarized as:

$$f_1 = -0.67Q_c + T_3 - 350 \leq 0$$

$$f_2 = -T_5 - 0.75T_1 + 0.5Q_c - T_3 + 1388.5 \leq 0$$

$$f_3 = -T_5 - 1.5T_1 + Q_c - 2T_3 + 2044 \leq 0 \qquad (3.23)$$

$$f_4 = -T_5 - 1.5T_1 + Q_c - 2T_3 - 2T_8 + 2830 \leq 0$$

$$f_5 = T_5 + 1.5T_1 - Q_c + 2T_3 + 3T_8 - 3153 \leq 0$$

where T_1, T_3, T_5, and T_8 are uncertain parameters that denote the fluid temperatures at various locations in the network, and Q_c is a controllable load in the cooler. The nominal values of these four temperatures are chosen at 620 K, 388 K, 583 K, and 313 K, respectively, in this example, whereas their expected positive and negative deviations are all set to be 10 K. Because there are four uncertain parameters in this case, the feasible region defined by Equation 3.23 cannot be actually visualized with a 4-D plot.

Notice first that Lai and Hui (2008) have already studied this problem and produced the following results:

- Using the auxiliary vectors, they mapped every uncertain parameter to a standard interval of $[-1,+1]$ and found that the volume estimate of the normalized feasible region was 12.14 units and the corresponding FI_v was 0.76. It should be noted that the true volume can be calculated by multiplying the scale factors, that is, $12.14 \times 10^4 = 121400$ units.

- Using the subspace feasibility tests, they divided the 4-D hypercube bounded by the expected upper and lower limits of uncertain parameters into 10,000 equal-sized hypercubic subspaces. The center of every subspace was then checked for feasibility. The flexibility index was calculated according to the following formula:

$$FI_v \approx \frac{N_{fs}}{N_s} \qquad (3.24)$$

- where N_{fs} is the number of feasible subspaces and N_s is the total number of subspaces (which is 10,000 in this example). Note that this approach may require overwhelming computation resources and may lead to over/under-estimation because the feasibility test is applied only to the center of each subspace. For the present example, FI_v was found to be 0.78.

The hypervolume of the feasible region has also been estimated repeatedly according to the proposed computation procedure with different numbers of boundary points. These estimates are plotted in Figure 3.20 for 10^3 to 10^5 points. It can be observed that the estimated hypervolume starts to stabilize after increasing the point number to a value higher than 5×10^4. The hypervolume obtained with 5×10^4 and 10^5 points can be found to be 126623.13 and 126742.43 units, respectively, whereas the corresponding flexibility indices in both cases are almost the same: 0.79. Finally, note that the hypervolume at 10^4 points (122434.22 units) is in fact quite close to the converged value, and thus, the corresponding flexibility index (0.77) can be used as a rough estimate for preliminary analysis.

3.3.6 HIGHER-DIMENSIONAL REGIONS

To show the superior capability of the proposed strategy in computing the hypervolumes of high-dimensional feasible regions, two examples are presented as follows:

1. A flow problem with five uncertain parameters (Grossmann and Floudas, 1987; Lai and Hui, 2008)
2. An HEN design problem with seven uncertain temperatures (Grossmann and Floudas, 1987; Lai and Hui, 2008)

For the sake of brevity, only the final results obtained with two existing approaches—auxiliary vector and subspace feasibility test—and the proposed computation strategy are presented in Table 3.1.

From these results one can see that

- The auxiliary vector approach usually underestimates the hypervolume of the feasible region volume.
- The subspace feasibility test is dependable but inefficient because all subspaces in the entire parameter hypercube have to be checked exhaustively.
- The proposed method may be adopted to produce accurate estimates of V_{fr} and FI_v with reasonable computation effort.

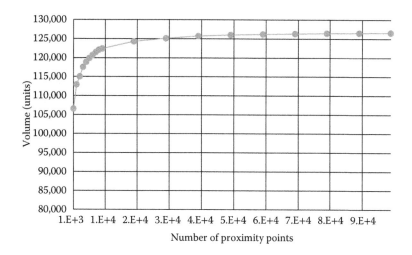

FIGURE 3.20 Effects of increasing boundary points in the heat exchanger network example. (Reprinted from *Chemical Engineering Science*, 147, Adi, V.S.K. et al., An effective computation strategy for assessing operational flexibility of high-dimensional systems with complicated feasible regions, 137–149. Copyright 2016, with permission from Elsevier, Ltd.).

TABLE 3.1

Computation Results in Higher-Dimensional Cases

Example		Auxiliary Vector (Lai and Hui, 2008)	Subspace Feasibility Test (Lai and Hui, 2008)	Proposed (10^n points)
Flow problem with five	V_{fr}	23.81	29.22	29.30
uncertain parameters	FI_v	0.744	0.913	0.92
HEN problem with seven	V_{fr}	118.4	126.72	126.22
uncertain parameters	FI_v	0.925	0.99	0.99

Source: Reprinted from *Chemical Engineering Science*, 147, Adi, V.S.K. et al., An effective computation strategy for assessing operational flexibility of high-dimensional systems with complicated feasible regions, 137–149. Copyright 2016, with permission from Elsevier, Ltd.

3.4 SUMMARY

An effective computation strategy is presented in this chapter for evaluating the volumetric flexibility index of high-dimensional systems with enhanced accuracy. The random nature of the line search provides a more precise characterization of the feasible region without *a priori* information of its geometric properties. By Delaunay triangulation, the hypervolumes of disjoint nonsimply connected and/or nonconvex regions can be computed accurately and efficiently. A heuristic rule is also suggested to facilitate proper selection of the number of boundary points. Finally, the effectiveness of the proposed computation strategy has been clearly demonstrated in a series of simple examples.

REFERENCES

Adi, V.S.K., Laxmidewi, R., Chang, C.T., 2016. An effective computation strategy for assessing operational flexibility of high-dimensional systems with complicated feasible regions. *Chemical Engineering Science* 147, 137–149.

Banerjee, I., Ierapetritou, M.G., 2002. Design optimization under parameter uncertainty for general black-box models. *Industrial & Engineering Chemistry Research* 41, 6687–6697.

Banerjee, I., Ierapetritou, M.G., 2003. Parametric process synthesis for general nonlinear models. *Computers & Chemical Engineering* 27, 1499–1512.

Banerjee, I., Ierapetritou, M.G., 2005. Feasibility evaluation of nonconvex systems using shape reconstruction techniques. *Industrial & Engineering Chemistry Research* 44, 3638–3647.

Banerjee, I., Pal, S., Maiti, S., 2010. Computationally efficient black-box modeling for feasibility analysis. *Computers & Chemical Engineering* 34, 1515–1521.

Barber, C.B., Dobkin, D.P., Huhdanpaa, H., 1996. The Quickhull algorithm for convex hulls. *ACM Transactions on Mathematical Software* 22, 469–483.

Bates, R.A., Wynn, H.R., Fraga, E.S., 2007. Feasible region approximation: A comparison of search cone and convex hull methods. *Engineering Optimization* 39, 513–527.

Boukouvala, F., Ierapetritou, M.G., 2012. Feasibility analysis of black-box processes using an adaptive sampling Kriging-based method. *Computers & Chemical Engineering* 36, 358–368.

Burkardt, J., 2013. Geometry: Geometric calculations. Department of Scientific Computing, Florida State University, Florida.

Croft, H.T., Falconer, K.J., Guy, R.K., 1994. *Unsolved Problems in Geometry*, 2nd corr. print. ed. New York, NY: Springer-Verlag.

Dirkse, S., Ferris, M.C., Ramakrishnan, J., 2014. *GDXMR—Interfacing GAMS and MATLAB*. Washington, DC: GAMS Development Corporation.

Goyal, V., Ierapetritou, M.G., 2003. Framework for evaluating the feasibility/operability of nonconvex processes. *Aiche Journal* 49, 1233–1240.

Grossmann, I.E., Floudas, C.A., 1987. Active constraint strategy for flexibility analysis in chemical processes. *Computers & Chemical Engineering* 11, 675–693.

Johnson, R.A., 2007. *Advanced Euclidean Geometry*. Mineola, NY: Dover Publications.

Klaus, S., 2010. The solid trefoil knot as an algebraic surface. *CIM Bulletin*, 228–4.

Krantz, S.G., 1999. *Handbook of Complex Variables*. Boston, MA: Birkhäuser.

Lai, S.M., Hui, C.W., 2008. Process flexibility for multivariable systems. *Industrial & Engineering Chemistry Research* 47, 4170–4183.

MathWorks, 2016a. *alphaShape*. The MathWorks, Inc., Massachusetts.

MathWorks, 2016b. *criticalAlpha*. The MathWorks, Inc., Massachusetts.

MathWorks, 2016c. *Delaunay triangulation*. The MathWorks, Inc., Massachusetts.

MathWorks, 2016d. *delaunayn*. The MathWorks, Inc., Massachusetts.

MathWorks, 2016e. *The language of technical computing*. The MathWorks, Inc., Massachusetts.

Rogers, A., Ierapetritou, M., 2015a. Feasibility and flexibility analysis of black-box processes Part 1: Surrogate-based feasibility analysis. *Chemical Engineering Science* 137, 986–1004.

Rogers, A., Ierapetritou, M., 2015b. Feasibility and flexibility analysis of black-box processes part 2: Surrogate-based flexibility analysis. *Chemical Engineering Science* 137, 1005–1013.

Rosenthal, R.E., 2016. *GAMS—A User's Guide GAMS Development Corporation*, Washington, DC: GAMS Development Corporation.

Stein, P., 1966. A note on the volume of a simplex. *American Mathematical Monthly* 73, 299–301.

Zilinskas, A., Fraga, E.S., Mackute, A., 2006. Data analysis and visualisation for robust multi-criteria process optimisation. *Computers & Chemical Engineering* 30, 1061–1071.

4 Dynamic Flexibility Index

As suggested by Dimitriadis and Pistikopoulos (1995), the operational flexibility of a *dynamic* system should be evaluated differently. By adopting a system of differential algebraic equations (DAEs) as the model constraints, these authors developed a mathematical programming formulation for dynamic flexibility analysis. Clearly, this analysis is more rigorous than that based on the steady-state model because, even for a continuous process, the operational flexibility cannot be adequately characterized without accounting for the control dynamics. In an earlier study, Brengel and Seider (1992) advocated the need for design and control integration. The integration of flexibility and controllability in design considerations was discussed extensively by several other groups (Aziz and Mujtaba, 2002; Bahri et al., 1997; Bansal et al., 1998; Chacon-Mondragon and Himmelblau, 1996; Georgiadis and Pistikopoulos, 1999; Malcolm et al., 2007; Mohideen et al., 1996a, b, 1997). Soroush and Kravaris (1993a, b) addressed various issues concerning flexible operation for batch reactors. The effects of uncertain disturbances on the wastewater neutralization processes were also studied by Walsh and Perkins (1994). White et al. (1996) presented an evaluation method to assess the switchability of any given system, that is, its ability to perform when moving between different operating points satisfactorily. Dimitriadis et al. (1997) studied the feasibility problem from the safety verification point of view. Zhou et al. (2009) utilized a similar approach to assess the operational flexibility of batch systems.

4.1 MODEL FORMULATION

For the dynamic flexibility analysis, the equality constraints in Equation 2.3 should be replaced with a system of DAEs (Dimitriadis and Pistikopoulos, 1995), which can be expressed in general form as

$$h_i\left(\mathbf{d}, \mathbf{z}(t), \mathbf{x}(t), \dot{\mathbf{x}}(t), \boldsymbol{\theta}(t)\right) = 0 \qquad \forall i \in \mathbb{I} \tag{4.1}$$

where $t \in [0, H]$, $i \in \mathbb{I}$, and $\mathbf{x}(0) = \mathbf{x}^0$. Also, the inequality constraints in Equation 2.4 must be rewritten accordingly:

$$g_j\left(\mathbf{d}, \mathbf{z}(t), \mathbf{x}(t), \boldsymbol{\theta}(t)\right) \leq 0 \qquad \forall j \in \mathbb{J} \tag{4.2}$$

Finally, the uncertain parameters and their upper and lower limits in this case should be regarded as functions of time, and thus, Equation 2.8 can be modified as follows:

$$\boldsymbol{\theta}^N(t) - \Delta\boldsymbol{\theta}^-(t) \leq \boldsymbol{\theta}(t) \leq \boldsymbol{\theta}^N(t) + \Delta\boldsymbol{\theta}^+(t) \tag{4.3}$$

Note that the time functions $\theta^N(t)$, $\Delta\theta^-(t)$, and $\Delta\theta^+(t)$ are assumed to be given *a priori*, and they may be extracted from historical records of observable time-dependent parameters (such as the hourly rainfall data collected every day during a period of several months or even years).

The corresponding *dynamic flexibility index FI_d* can be computed on the basis of the following model:

$$FI_d = \max \; \delta \qquad (4.4)$$

subject to Equation 4.1 and

$$\max_{\theta(t)} \min_{z(t),x(t)} \max_{j,t} g_j\big(d,z(t),x(t),\theta(t)\big) \le 0 \qquad (4.5)$$

$$\theta^N(t) - \delta\Delta\theta^-(t) \le \theta(t) \le \theta^N(t) + \delta\Delta\theta^+(t) \qquad (4.6)$$

Note that this model is essentially the dynamic version of Equations 2.11 and 2.12.

4.2 NUMERICAL SOLUTION STRATEGIES

Two alternative solution strategies are presented in this section for computing the dynamic flexibility index. First of all, Equations 4.1 and 4.4 through 4.6 can obviously be transformed into a steady-state flexibility index model by approximating the embedded differential equations with a set of algebraic equations. The vertex method described in Section 2.2.2 is readily applicable for solving this transformed model. Another viable option is to establish the Karush-Kuhn-Tucker (KKT) conditions for the aforementioned dynamic programming model and then develop the corresponding active set method. Before illustrating these approaches in detail, let us first briefly outline two alternative discretization techniques.

4.2.1 DISCRETIZATION OF DYNAMIC MODEL

An obvious solution approach for computing the dynamic flexibility index is to first convert the nonlinear DAEs in Equation 4.1 into a system of algebraic equations by a credible numerical discretization technique. Although many equally effective techniques are available, only two of them—the differential quadrature (DQ) (Bellman and Casti, 1971; Bellman et al., 1972; Naadimuthu et al., 1984) and the trapezoidal rule (TR)—are provided in the sequel to facilitate clear explanation.

4.2.1.1 Differential Quadrature (DQ)

The accuracy of DQ approximation has been well documented in the literature (Chang et al., 1993; Civan and Sliepcevich, 1984; Quan and Chang, 1989a, b), and its implementation procedure is very straightforward. As pointed out by Shu (2000), DQ is essentially equivalent to the finite difference scheme of a higher order. To improve the computation efficiency when a large number of grid points are required, a localized DQ scheme was adopted by Zong and Lam (2002). An extensive discussion of DQ and its state-of-the-art developments can be found in Zong and Zhang (2009).

To illustrate the DQ discretization principle, let us consider the first-order derivative of the ith state variable ($i \in \mathbb{I}$) as an example:

$$\left. \frac{dx_i(t)}{dt} \right|_{t=t_m^e} \cong \sum_{n=1}^{N_{\text{node}}} w_{mn} x_i \left(t_n^e \right) \qquad (4.7)$$

where $m \in \{1, 2, \cdots, N_{\text{node}}\}$; $e \in \{1, 2, \cdots, N_{\text{element}}\}$; t_m^e and t_n^e, respectively, denote the locations of the mth and n^{th} nodes in time element e; and w_{mn} denotes the weighting coefficient associated with the state value at t_n^e for the derivative at t_m^e, which is dependent only upon a *predetermined* node spacing. As a result, every differential equation in Equation 4.1 can be approximated with a set of algebraic equations. In addition, all inequalities in Equations 4.2 should be discretized at the node locations, that is,

$$g_j \left(\mathbf{d}, \mathbf{z}\left(t_m^e \right), \mathbf{x}\left(t_m^e \right), \boldsymbol{\theta}\left(t_m^e \right) \right) \le 0 \qquad \forall j \in \mathbb{J} \qquad (4.8)$$

Quan and Chang (1989a, b) suggested that, in most cases, it is beneficial to use the shifted zeros of a standard Chebyshev polynomial as the selected nodes. This node spacing in an arbitrary interval $t \in [a,b]$ yields the following formulas for calculating the weighting coefficients:

$$w_{mn} = \frac{r_{N_{\text{node}}} - r_1}{b-a} \frac{(-1)^{(m-n)}}{r_m - r_n} \sqrt{\frac{1-r_n^2}{1-r_m^2}}, \qquad m \ne n \qquad (4.9)$$

$$w_{mm} = \frac{1}{2} \frac{r_{N_{\text{node}}} - r_1}{b-a} \frac{r_m}{(1-r_m^2)} \qquad (4.10)$$

where $m, n \in \{1, 2, \cdots, N_{\text{node}}\}$ and the locations of Chebyshev zeros in the standard interval $[-1, +1]$ are

$$r_m = \cos \frac{(2m-1)\pi}{2N_{\text{node}}} \qquad (4.11)$$

where w_{mn} are the weighting coefficients for the first-order derivatives. With these formulas, all weighting coefficients can be easily calculated for any combination of element length $b-a$ and node number N_{node}. A typical example can be found in Table 4.1.

The time horizon $[0, H]$ is supposed to be divided into N_{element} elements as mentioned previously. Continuity of every state variable at the border point of each pair of adjacent elements can be enforced with a boundary condition, that is, $\mathbf{x}\left(t_{N_{\text{node}}}^e \right) = \mathbf{x}\left(t_1^{e+1} \right)$, and $e \in \{1, 2, \cdots, N_{\text{element}} - 1\}$. The initial conditions of Equation 4.1 should be imposed at the left end of the first element, that is, $\mathbf{x}\left(t_1^1 \right) = \mathbf{x}^0$, whereas the

TABLE 4.1

Weighting Coefficients for $b - a = 10$ and $N_{node} = 5$

m n	1	2	3	4	5
1	0.90085	−0.2	0.06180	0.04721	0.1
2	1.37082	0.05020	−0.26180	0.16180	−0.32361
3	−0.64721	0.4	0	−0.4	0.64721
4	0.32361	−0.16180	0.26180	0.05020	−1.37082
5	−0.1	0.04721	−0.06180	0.2	0.90085

Source: Reprinted with permission from Kuo (2015). Copyright 2015 National Cheng Kung University Library.

states at the right end of the last element are not constrained, that is, $\mathbf{x}\left(t_{N_{node}}^{N_{element}}\right) =$ free. Finally, the element number and lengths should be allowed to be adjusted to achieve satisfactory accuracy.

4.2.1.2 Trapezoidal Rule (TZ)

For illustration clarity, let us first rewrite Equation 4.1 as follows:

$$\frac{d\mathbf{x}(t)}{dt} = \boldsymbol{\varphi}\left(\mathbf{d}, \mathbf{x}(t), \mathbf{z}(t), \boldsymbol{\theta}(t)\right) \tag{4.12}$$

or

$$\frac{dx_i(t)}{dt} = \varphi_i\left(\mathbf{d}, \mathbf{x}(t), \mathbf{z}(t), \boldsymbol{\theta}(t)\right), \quad i \in \mathbb{I} \tag{4.13}$$

Let us divide the horizon $[0, H]$ into M equal intervals and label their boundary points sequentially as $p = 0, 1, 2, \cdots, M$. Thus, the length of each interval should be

$$h = \frac{H}{M} \tag{4.14}$$

By applying the TR to estimate the integral of $\varphi_i\left(\mathbf{d}, \mathbf{x}(t), \mathbf{z}(t), \boldsymbol{\theta}(t)\right)$ over each interval, one can obtain

$$x_i\left(t_p\right) = x_i\left(t_{p-1}\right) + \frac{h}{2}\left[\varphi_i\left(\mathbf{d}, \mathbf{x}\left(t_{p-1}\right), \mathbf{z}\left(t_{p-1}\right), \boldsymbol{\theta}\left(t_{p-1}\right)\right) + \varphi_i\left(\mathbf{d}, \mathbf{x}\left(t_p\right), \mathbf{z}\left(t_p\right), \boldsymbol{\theta}\left(t_p\right)\right)\right] \tag{4.15}$$

where $x_i(0) = x_i^0$, $i \in \mathbb{I}$, and $p = 1, 2, \cdots, M$. Similarly, Equation 4.2 can also be discretized according to the aforementioned boundary points as follows:

$$g_j\left(\mathbf{d}, \mathbf{x}\left(t_p\right), \mathbf{z}\left(t_p\right), \boldsymbol{\theta}\left(t_p\right)\right) \le 0 \quad \forall j \in \mathbb{J} \tag{4.16}$$

4.2.2   EXTENDED VERTEX METHOD—DYNAMIC VERSION

The dynamic version of the vertex method (Kuo and Chang, 2016) can be formulated as a two-step optimization problem, that is,

$$FI_d = \min_{k} \max_{\mathbf{z}(t),\mathbf{x}(t)} \delta \tag{4.17}$$

subject to Equations 4.1 and 4.2 and the following vertex constraint in the functional space formed by all possible $\theta(t)$:

$$\theta(t) = \theta^N(t) + \delta\Delta\theta^k(t) \tag{4.18}$$

where $\Delta\theta^k(t)$ is a vector that points from the nominal point $\theta^N(t)$ toward the kth vertex ($k = 1,2,\cdots,2^{n_p}$ and n_p is the number of uncertain parameters) at time t. Note that each element in $\Delta\theta^k(t)$ should be obtained from the corresponding entry in either $-\Delta\theta^-(t)$ or $+\Delta\theta^+(t)$.

For illustration clarity, let us next produce a specific formulation by discretizing the previous model with the TR:

$$FI_d = \min_{k} \max_{\mathbf{Z},\mathbf{X}} \delta \tag{4.19}$$

subject to Equations 4.14 through 4.16 and

$$\theta(t_p) = \theta^N(t_p) + \delta\Delta\theta^k(t_p), \quad p = 1,2,...,M \tag{4.20}$$

where $X = \begin{bmatrix} \mathbf{x}(t_1) & \mathbf{x}(t_2) & \cdots & \mathbf{x}(t_M) \end{bmatrix}$ and $Z = \begin{bmatrix} \mathbf{z}(t_1) & \mathbf{z}(t_2) & \cdots & \mathbf{z}(t_M) \end{bmatrix}$.

Figure 4.1 shows a flowchart of the computation procedure that realizes this version of the extended vertex method.

4.2.3   EXTENDED ACTIVE SET METHOD—DYNAMIC VERSION

A dynamic version of the feasibility function (or the feasibility functional) proposed by Wu and Chang (2017) can be defined in the same way as its steady-state counterpart in Equations 2.3, 2.6, and 2.7, that is,

$$\psi(\mathbf{d},\theta(t)) = \min_{\mathbf{x}(t),\mathbf{z}(t),u(t)} u(t)\big|_{t=H} \tag{4.21}$$

subject to the constraints in Equation 4.1 and

$$\dot{u}(t) = 0 \tag{4.22}$$

$$g_j(\mathbf{d},\mathbf{z}(t),\mathbf{x}(t),\theta(t)) \leq u(t) \quad \forall j \in \mathbb{J} \tag{4.23}$$

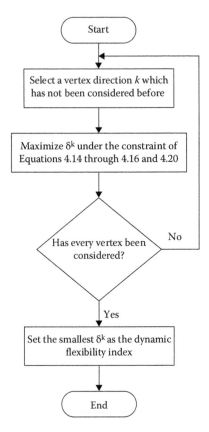

FIGURE 4.1 The dynamic version of the extended vertex method (Reprinted with permission from Kuo [2015] Copyright 2015 National Cheng Kung University Library.)

To facilitate derivation of the KKT conditions for this functional optimization problem, let us express Equation 4.1 regarding Equation 4.12, that is,

$$\mathbf{h}\big(\mathbf{d},\mathbf{z}(t),\mathbf{x}(t),\dot{\mathbf{x}}(t),\theta(t)\big) = \varphi\big(\mathbf{d},\mathbf{z}(t),\mathbf{x}(t),\theta(t)\big) - \dot{\mathbf{x}}(t) = 0 \qquad (4.24)$$

An aggregated objective functional can then be constructed by introducing Lagrange multipliers to incorporate all constraints:

$$L = u(H) + \int_0^H \left\{ \mu_u(t)[0-\dot{u}] + \mu(t)^T\big[\varphi - \dot{\mathbf{x}}\big] + \lambda(t)^T\big[\mathbf{g} - u\mathbf{1}\big] \right\} dt \qquad (4.25)$$

where, $\mathbf{g} = \begin{bmatrix} g_1 & g_2 & g_3 \cdots \end{bmatrix}^T$ and $\mathbf{1} = \begin{bmatrix} 1 & 1 & 1 \cdots \end{bmatrix}^T$. Note that the multipliers $\mu_u(t)$ and $\mu(t)$ are real, but $\lambda(t) \geq 0$. By taking the first variation of L and setting it to zero, the following four sets of necessary conditions can be obtained:

1. $\mathbf{x}(0) = \mathbf{x}_0;\ \mu(H) = \mathbf{0};\ \mu_u(0) = 0;\ \mu_u(H) = 1$

2. $\dot{\mu} = -\mu^T\left(\dfrac{\partial \varphi}{\partial \mathbf{x}}\right) - \lambda^T\left(\dfrac{\partial \mathbf{g}}{\partial \mathbf{x}}\right);\ \dot{\mu}_u = \lambda^T\mathbf{1}$

3. $\dot{\mathbf{x}} = \boldsymbol{\varphi}$; $\dot{\mathbf{u}} = 0$; $\boldsymbol{\lambda}^T(\mathbf{g} - u\mathbf{1}) = 0$; $\boldsymbol{\lambda} \geq \mathbf{0}$

4. $\boldsymbol{\mu}^T\left(\dfrac{\partial\boldsymbol{\varphi}}{\partial\mathbf{z}}\right) + \boldsymbol{\lambda}^T\left(\dfrac{\partial\mathbf{g}}{\partial\mathbf{z}}\right) = \mathbf{0}$

By following the same rationale in developing the active set method for computing the steady-state flexibility index FI_s, that is, Equations 2.13 and 2.14, it is necessary to set $u(t) = 0$ and change conditions in (iii) to

$$\text{(iii)}' \quad \dot{\mathbf{x}} = \boldsymbol{\varphi}; \; u = 0; \; \boldsymbol{\lambda}^T\mathbf{g} = 0; \; \boldsymbol{\lambda} \geq \mathbf{0}; \; \mathbf{g} \leq \mathbf{0}$$

Therefore, the dynamic flexibility index FI_d can be determined by minimizing δ while subject to the constraints in necessary conditions (i), (ii), (iii)', (iv), and Equation 4.6.

To facilitate practical implementation of the previous ideas, additional slack and binary variables should be introduced to reformulate the last three conditions in (iii)', that is, $\boldsymbol{\lambda}^T\mathbf{g} = 0$, $\boldsymbol{\lambda} \geq \mathbf{0}$, and $\mathbf{g} \leq \mathbf{0}$. Specifically, the optimization problem earlier can be written as

$$FI_d = \min_{\delta,\mu_u(t),\mu_i(t),\lambda_j(t),s_j(t),y_j(t),x_i(t),z_c(t),\theta_n(t)} \delta \qquad (4.26)$$

subject to

$$\frac{dx_i(t)}{dt} = \varphi_i\big(\mathbf{d},\mathbf{x}(t),\mathbf{z}(t),\boldsymbol{\theta}(t)\big), \quad x_i(0) = x_i^0, \quad \forall i \in \mathbb{I}; \qquad (4.27)$$

$$g_j\big(\mathbf{d},\mathbf{z}(t),\mathbf{x}(t),\boldsymbol{\theta}(t)\big) + s_j(t) = 0, \quad s_j(t) \geq 0, \quad \forall j \in \mathbb{J}; \qquad (4.28)$$

$$\dot{\mu}_u(t) = \sum_{j'\in\mathbb{J}} \lambda_{j'}(t), \quad \mu_u(0) = 0, \quad \mu_u(H) = 1; \qquad (4.29)$$

$$\dot{\mu}_i(t) = -\sum_{i'\in\mathbb{I}} \mu_{i'}(t)\frac{\partial\varphi_{i'}}{\partial x_i}(t) - \sum_{j'\in\mathbb{J}} \lambda_{j'}(t)\frac{\partial g_{j'}}{\partial x_i}(t), \quad \mu_i(H) = 0, \quad \forall i \in \mathbb{I}; \qquad (4.30)$$

$$\sum_{i'\in\mathbb{I}} \mu_{i'}(t)\frac{\partial\varphi_{i'}}{\partial z_c}(t) + \sum_{j'\in\mathbb{J}} \lambda_{j'}(t)\frac{\partial g_{j'}}{\partial z_c}(t) = 0, \quad c = 1,2,\cdots,n_z; \qquad (4.31)$$

$$s_j(t) - U\big(1 - y_j(t)\big) \leq 0, \; \lambda_j(t) - y_j(t) \leq 0, \; y_j(t) \in \{0,1\}, \; \lambda_j(t) \geq 0, \; \forall j \in \mathbb{J}; \qquad (4.32)$$

$$\theta_n^N(t) - \delta\Delta\theta_n^-(t) \leq \theta_n(t) \leq \theta_n^N(t) + \delta\Delta\theta_n^+(t), \quad n = 1,2,\cdots,n_p. \qquad (4.33)$$

where $0 < t \leq H$. An implementable formulation can then be produced by discretizing Equations 4.28 through 4.35 with the TR. Notice that, unlike the exhaustive enumeration approach described in Figure 4.1, it is only necessary to solve the resulting mixed integer nonlinear programming (MINLP) problem once.

4.3 AN ILLUSTRATIVE EXAMPLE

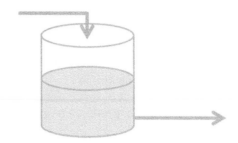

FIGURE 4.2 A buffer vessel. (Reprinted with permission from Kuo and Chang, 2016, 670–682. Copyright 2016 American Chemical Society.)

Let us consider the buffer tank in Figure 4.2. The dynamic model of this system can be written as

$$A\frac{dh}{dt} = \theta(t) - k\sqrt{h} \tag{4.34}$$

where h denotes the height of liquid level (m); A ($= 5$ m^2) is the cross-sectional area of the tank; k ($= \sqrt{5}/10$ m$^{5/2}$ min^{-1}) is a proportionality constant; and θ denotes the feed flow rate (m^3 min^{-1}), and it is treated as the only uncertain parameter in the present example. To fix the ideas, the following upper and lower limits are also adopted in the flexibility analysis:

1. The height of the tank is 10 m, that is, $h \le 10$.
2. Due to the operational requirement of downstream unit(s), the outlet flow rate of the buffer tank must be kept above $\dfrac{\sqrt{5}}{10}$ m^3 min^{-1}. Thus, the minimum allowable height of its liquid level should be 1 m, that is, $1 \le h$.
3. The time horizon covers a period of 800 minutes. In other words, this interval can be expressed as $0 \le t \le 800$.

To interpret the dynamic flexibility index, two different operation modes are considered in the sequel.

Case 4.1: Continuous Operation

Let us assume that in the buffer operation under consideration, the nominal steady-state value of the feed rate is $\theta^N(t) = 0.5$ m^{3}min^{-1} and the corresponding anticipated positive and negative deviations are set at $\Delta\theta^+(t) = \Delta\theta^-(t) = 0.5$ m^{3}min^{-1}, respectively. Therefore, the range of the uncertain parameter is

$$0 \le \theta(t) \le 1 \tag{4.35}$$

and the nominal height of liquid level at steady state should be 5 m. By discretizing Equation 4.34 according to the TR (see Section 4.2.1.2) and then applying the dynamic version of the extended vertex method (see Section 4.2.2), one can evaluate the corresponding dynamic flexibility index, and its value is 0.415. To facilitate further clarification, the GAMS code of this computation procedure is provided in Appendix 4.1.

This result indicates that although the present system is not operable in the worst-case scenario, FI_d can be raised to the desired target of 1 if the expected range of the uncertain parameter can be narrowed by improving the flow control quality of the feed stream to

$$0.5 - 0.415 \times 0.5 \leq \theta(t) \leq 0.5 + 0.415 \times 0.5 \tag{4.36}$$

This assertion can be verified by carrying out numerical simulations of the worst-case scenarios (see Figure 4.3). One can observe that if the feed rate is maintained, respectively, at the upper and lower limits of the narrowed range defined in Equation 4.36, the water level can be guaranteed to satisfy the operational constraints at any time throughout the given horizon. Furthermore, one can also observe that (1) the water level approaches 10 m (i.e., the upper bound of h) at 800 min in the former scenario when $\theta(t) = 0.7075$ and (2) the water level always stays considerably above the lower limit (i.e., $h > 1$) in the latter case when $\theta(t) = 0.2925$.

Note finally that if it is not possible to improve the control quality of the upstream feed stream, the operational target of $FI_d = 1$ can be realized alternatively by increasing the buffer capacity. In particular, a larger storage tank with a cross-sectional area of 61 m² can be adopted to replace the original one to withstand all possible disturbances constrained by Equation 4.35.

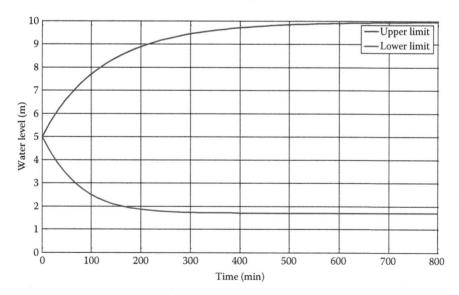

FIGURE 4.3 Simulation results of the worst-case scenarios for the narrowed parameter range defined in Equation 4.28 in Case 4.1. (Reprinted with permission from Kuo and Chang, 2016, 670–682. Copyright 2016 American Chemical Society.)

Case 4.2: Periodic Operation

Let us assume that over a single period of 800 minutes in the cyclic operation under consideration, the nominal feed rate and its anticipated positive and negative deviations can be described in Equation 4.37 and in Figure 4.4. The initial height of the liquid level in this case is also set at 5 m.

$$
\begin{cases}
\theta^N(t) = 0.5 \ (\text{m}^3\,\text{min}^{-1}), \ \Delta\theta^+(t) = \Delta\theta^-(t) = 0.1 \quad \text{for } 0 \le t \le 100 \ (\text{min}) \\[4pt]
\theta^N(t) = 0.6 \ (\text{m}^3\,\text{min}^{-1}), \ \Delta\theta^+(t) = \Delta\theta^-(t) = 0.1 \quad \text{for } 100 \le t \le 200 \ (\text{min}) \\[4pt]
\theta^N(t) = 0.7 \ (\text{m}^3\,\text{min}^{-1}), \ \Delta\theta^+(t) = \Delta\theta^-(t) = 0.1 \quad \text{for } 200 \le t \le 250 \ (\text{min}) \\[4pt]
\theta^N(t) = 0.8 \ (\text{m}^3\,\text{min}^{-1}), \ \Delta\theta^+(t) = \Delta\theta^-(t) = 0.1 \quad \text{for } 250 \le t \le 300 \ (\text{min}) \\[4pt]
\theta^N(t) = 0.6 \ (\text{m}^3\,\text{min}^{-1}), \ \Delta\theta^+(t) = \Delta\theta^-(t) = 0.1 \quad \text{for } 300 \le t \le 350 \ (\text{min}) \\[4pt]
\theta^N(t) = 0.5 \ (\text{m}^3\,\text{min}^{-1}), \ \Delta\theta^+(t) = \Delta\theta^-(t) = 0.1 \quad \text{for } 350 \le t \le 450 \ (\text{min}) \\[4pt]
\theta^N(t) = 0.4 \ (\text{m}^3\,\text{min}^{-1}), \ \Delta\theta^+(t) = \Delta\theta^-(t) = 0.1 \quad \text{for } 450 \le t \le 500 \ (\text{min}) \\[4pt]
\theta^N(t) = 0.2 \ (\text{m}^3\,\text{min}^{-1}), \ \Delta\theta^+(t) = \Delta\theta^-(t) = 0.1 \quad \text{for } 500 \le t \le 600 \ (\text{min}) \\[4pt]
\theta^N(t) = 0.6 \ (\text{m}^3\,\text{min}^{-1}), \ \Delta\theta^+(t) = \Delta\theta^-(t) = 0.1 \quad \text{for } 600 \le t \le 700 \ (\text{min}) \\[4pt]
\theta^N(t) = 0.5 \ (\text{m}^3\,\text{min}^{-1}), \ \Delta\theta^+(t) = \Delta\theta^-(t) = 0.1 \quad \text{for } 700 \le t \le 800 \ (\text{min})
\end{cases}
\tag{4.37}
$$

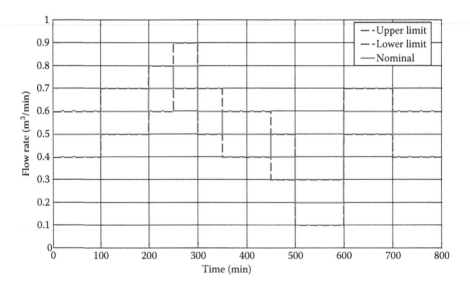

FIGURE 4.4 Nominal feed rate and its upper and lower limits in a single period in Case 4.2. (Reprinted with permission from Kuo and Chang, 2016, 670–682. Copyright 2016 American Chemical Society.)

By discretizing Equation 4.34 according to the TR (see Section 4.2.1.2) and then applying the dynamic vertex method (see Section 4.2.2), one can find that the corresponding dynamic flexibility index is 0.368 (see the GAMS code in Appendix 4.2). This index value indicates that although the present system is inoperable in the worst-case scenario, FI_d can be made to achieve the designated target of 1 by reducing the range of variation in the uncertain parameter, that is,

$$\theta^N(t) - 0.368 \times \Delta\theta^-(t) \le \theta(t) \le \theta^N(t) + 0.368 \times \Delta\theta^+(t) \qquad (4.38)$$

where $\theta^N(t)$, $\Delta\theta^-(t)$, and $\Delta\theta^+(t)$ are defined in Equation 4.37. Figure 4.5 shows the simulation results of the corresponding worst-case scenarios. Note that if the feed rate is maintained, respectively, at the upper and lower limits of the narrowed parameter range defined in Equation 4.38, the water level should stay within the allowed range—that is, $1 \le h \le 10$ —at any time throughout the given horizon. It can be observed that the water level reaches 1 m (which is the lower bound of h) at 600 min in the latter case. Obviously, the operational target of $FI_d = 1$ can also be achieved by enlarging the cross-sectional area of the buffer tank to 8.25 m². The simulation results of the corresponding worst-case scenarios can be found in Figure 4.6. Finally, it should be noted that the same FI_d (0.368) can be obtained in this case with the extended active set method. However, the needed computation time in GAMS 23.9.5 on an i7-4770 34 GHz PC was found to be 4 sec, which is slightly longer than that consumed with the extended vertex method (2 sec).

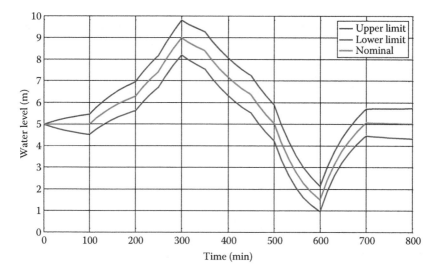

FIGURE 4.5 Simulation results of the worst-case scenarios for the narrowed parameter range defined in Equation 4.38 in Case 4.2. (Reprinted with permission from Kuo and Chang, 2016, 670–682. Copyright 2016 American Chemical Society.)

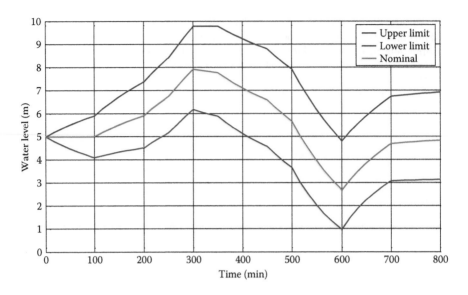

FIGURE 4.6 Simulation results of the worst-case scenarios for the original parameter range defined in Equation 4.37 and an enlarged buffer tank in Case 4.2. (Reprinted with permission from Kuo and Chang, 2016, 670–682. Copyright 2016 American Chemical Society.)

4.4 CUMULATIVE EFFECTS OVERLOOKED IN DYNAMIC FLEXIBILITY ANALYSIS

As mentioned in Section 4.1, the time-dependent range of an uncertain parameter is often extracted directly from historical data without elaborate statistical interpretations. However, the cumulated quantities of these parameters over time may also be recorded, and these available data are in general neglected in the aforementioned dynamic flexibility analysis. Specifically, in addition to Equation 4.3, the following extra inequalities can often be established to characterize the uncertain parameters better:

$$-\Delta\Theta^- \leq \Theta(H) \leq +\Delta\Theta^+ \tag{4.39}$$

where

$$\Theta(t) = \int^t \left[\theta(\tau) - \theta^N(\tau)\right] d\tau \tag{4.40}$$

As also indicated before, the time functions $\theta^N(t)$, $\Delta\theta^-(t)$, and $\Delta\theta^+(t)$ in Equation 4.3 can usually be extracted from historical records of transient data. To fix ideas, let us consider the precipitation data as an example. By setting H to be 24 hours, these time functions may be established on the basis of the *hourly* rainfall records collected every day over a period of several months, and, in addition, the scalar values of $\Delta\Theta^-$ and $\Delta\Theta^+$ in Equation 4.39 can be estimated according to the *daily* rainfall data, which are usually also available. Because the random deviations in parameters do not always stay at the upper or lower limits throughout the entire horizon $[0, H]$, one would expect

$$\Delta\Theta^- \leq \int_0^H \Delta\theta^-(\tau)d\tau \tag{4.41}$$

$$\Delta\Theta^+ \leq \int_0^H \Delta\theta^+(\tau)d\tau \tag{4.42}$$

Because the uncertainties in an unsteady system should be better modeled with both Equations 4.3 and 4.39, there are clear incentives to develop a different flexibility index accordingly.

APPENDIX 4.1:   GAMS CODE USED IN CASE 4.1 OF THE BUFFER TANK EXAMPLE

```
sets
m discrete time /1*800/
;
parameters
Ar area of tank /5/
k;
k=5**0.5/10;
positive variables
fl flexibility index
h(m)   tank level
the(m)   flow rate
;
variables
z
;
h.l(m)=5;
equations
eqcon(m),ineqcon1(m),ineqcon2(m),up(m),h0,min
;
eqcon(m)$(ord(m)<800)..Ar*(h(m+1)-h(m))=e=0.5*(the(m+1)+
the(m))-0.5*k*(h(m+1)**0.5+h(m)**0.5);
ineqcon1(m)..(h(m)-10)=l=0;
ineqcon2(m)..(1-h(m))=l=0;
up(m)..the(m)=e=0.5+0.5*fl;
h0..h('1')=e=5;
min..z=e=fl;
model aaa /all/;
solve aaa using nlp maximizing z;
display z.l,h.l,the.l;
```

APPENDIX 4.2:   GAMS CODE USED IN CASE 4.2 OF THE BUFFER TANK EXAMPLE

```
sets
m discrete time /1*800/;
parameters
Ar area of tank /5/
dt uncertain deviation /0.1/
k;
k=5**0.5/10;
positive variables
fl   flexibility index
h(m)   tank level
the(m) flow rate;
variables
z ;
h.l(m)=5;
equations
eqcon(m),ineqcon1(m),ineqcon2(m),  h0,min
up1(m),up2(m),up3(m),up4(m),up5(m),up6(m),up7(m),up8(m),up9(m),
up10(m) ;
eqcon(m)$(ord(m)<800)..Ar*(h(m+1)-h(m))=e=0.5*(the(m+1)+
the(m))-0.5*k*(h(m+1)**0.5+h(m)**0.5);
ineqcon1(m)..(h(m)-10)=l=0;
ineqcon2(m)..(1-h(m))=l=0;
up1(m)$(ord(m)<101)..the(m)=e=0.5+dt*fl;
up2(m)$(ord(m)<201 and ord(m)>100)..the(m)=e=0.6+dt*fl;
up3(m)$(ord(m)<251 and ord(m)>200)..the(m)=e=0.7+dt*fl;
up4(m)$(ord(m)<301 and ord(m)>250)..the(m)=e=0.8+dt*fl;
up5(m)$(ord(m)<351 and ord(m)>300)..the(m)=e=0.6+dt*fl;
up6(m)$(ord(m)<451 and ord(m)>350)..the(m)=e=0.5+dt*fl;
up7(m)$(ord(m)<501 and ord(m)>450)..the(m)=e=0.4+dt*fl;
up8(m)$(ord(m)<601 and ord(m)>500)..the(m)=e=0.2+dt*fl;
up9(m)$(ord(m)<701 and ord(m)>600)..the(m)=e=0.6+dt*fl;
up10(m)$(ord(m)>700)..the(m)=e=0.5+dt*fl;
h0..h('1')=e=5;
min..z=e=fl;
model aaa /all/;
solve aaa using minlp maximizing z;
display z.l,h.l,the.l;
```

REFERENCES

Aziz, N., Mujtaba, I.M., 2002. Optimal operation policies in batch reactors. *Chemical Engineering Journal* 85, 313–325.

Bahri, P.A., Bandoni, J.A., Romagnoli, J.A., 1997. Integrated flexibility and controllability analysis in design of chemical processes. *Aiche Journal* 43, 997–1015.

Bansal, V., Perkins, J.D., Pistikopoulos, E.N., 1998. Flexibility analysis and design of dynamic processes with stochastic parameters. *Computers & Chemical Engineering* 22, S817–S820.

Bellman, R., Casti, J., 1971. Differential quadrature and long-term integration. *Journal of Mathematical Analysis and Applications* 34, 235–238.

Bellman, R., Casti, J., Kashef, B.G., 1972. Differential quadrature - technique for rapid solution of nonlinear partial differential equations. *Journal of Computational Physics* 10, 40–52.

Brengel, D.D., Seider, W.D., 1992. Coordinated design and control optimization of nonlinear processes. *Computers & Chemical Engineering* 16, 861–886.

Chacon-Mondragon, O.L., Himmelblau, D.M., 1996. Integration of flexibility and control in process design. *Computers & Chemical Engineering* 20, 447–452.

Chang, C.T., Tsai, C.S., Lin, T.T., 1993. The modified differential quadratures and their applications. *Chemical Engineering Communications* 123, 135–164.

Civan, F., Sliepcevich, C.M., 1984. Differential quadrature for multi-dimensional problems. *Journal of Mathematical Analysis and Applications* 101, 423–443.

Dimitriadis, V.D., Pistikopoulos, E.N., 1995. Flexibility analysis of dynamic-systems. *Industrial & Engineering Chemistry Research* 34, 4451–4462.

Dimitriadis, V.D., Shah, N., Pantelides, C.C., 1997. Modeling and safety verification of discrete/continuous processing systems. *Aiche Journal* 43, 1041–1059.

Georgiadis, M.C., Pistikopoulos, E.N., 1999. An integrated framework for robust and flexible process systems. *Industrial & Engineering Chemistry Research* 38, 133–143.

Kuo, Y.C., 2015. Applications of the dynamic and temporal flexibility indices, Department of Chemical Engineering. National Cheng Kung University, Tainan.

Kuo, Y.C., Chang, C.T., 2016. On heuristic computation and application of flexibility indices for unsteady process design. *Industrial & Engineering Chemistry Research* 55, 670–682.

Malcolm, A., Polan, J., Zhang, L., Ogunnaike, B.A., Linninger, A.A., 2007. Integrating systems design and control using dynamic flexibility analysis. *Aiche Journal* 53, 2048–2061.

Mohideen, M.J., Perkins, J.D., Pistikopoulos, E.N., 1996a. Optimal design of dynamic systems under uncertainty. *Aiche Journal* 42, 2251–2272.

Mohideen, M.J., Perkins, J.D., Pistikopoulos, E.N., 1996b. Optimal synthesis and design of dynamic systems under uncertainty. *Computers & Chemical Engineering* 20, S895–S900.

Mohideen, M.J., Perkins, J.D., Pistikopoulos, E.N., 1997. Robust stability considerations in optimal design of dynamic systems under uncertainty. *Journal of Process Control* 7, 371–385.

Naadimuthu, G., Bellman, R., Wang, K.M., Lee, E.S., 1984. Differential quadrature and partial-differential equations - Some numerical results. *Journal of Mathematical Analysis and Applications* 98, 220–235.

Quan, J.R., Chang, C.T., 1989a. New insights in solving distributed system equations by the quadrature method .1. Analysis. *Computers & Chemical Engineering* 13, 779–788.

Quan, J.R., Chang, C.T., 1989b. New insights in solving distributed system equations by the quadrature method .2. Numerical experiments. *Computers & Chemical Engineering* 13, 1017–1024.

Shu, C., 2000. Differential quadrature and its application in engineering. Springer London.

Soroush, M., Kravaris, C., 1993a. Optimal-design and operation of batch reactors.1. Theoretical framework. *Industrial & Engineering Chemistry Research* 32, 866–881.

Soroush, M., Kravaris, C., 1993b. Optimal-design and operation of batch reactors.2. A case-study. *Industrial & Engineering Chemistry Research* 32, 882–893.

Walsh, S., Perkins, J., 1994. Application of integrated process and control-system design to waste-water neutralization. *Computers & Chemical Engineering* 18, S183–S187.

White, V., Perkins, J.D., Espie, D.M., 1996. Switchability analysis. *Computers & Chemical Engineering* 20, 469–474.

Wu, R.S., Chang, C.T., 2017. Development of mathematical programs for evaluating dynamic and temporal flexibility indices based on KKT conditions. *Journal of the Taiwan Institute of Chemical Engineers.* 73, 86–92.

Zhou, H., Li, X.X., Qian, Y., Chen, Y., Kraslawski, A., 2009. Optimizing the initial conditions to improve the dynamic flexibility of batch processes. *Industrial & Engineering Chemistry Research* 48, 6321–6326.

Zong, Z., Lam, K.Y., 2002. A localized differential quadrature (LDQ) method and its application to the 2D wave equation. *Computational Mechanics* 29, 382–391.

Zong, Z., Zhang, Y., 2009. *Advanced Differential Quadrature Methods, Chapman & Hall/ CRC Applied Mathematics and Nonlinear Science Series.* Boca Raton, FL: CRC Press, p. 1 online resource (339 p.).

5 Temporal Flexibility Index

As mentioned in Section 4.4, there are real incentives to develop a quantitative flexibility measure that takes into account the accumulated effects of uncertain parameters in nonsteady processes. For this purpose, Adi and Chang (2013) suggested computing a so-called *temporal flexibility index* (FI_t) by solving an optimization problem similar to that described in the previous chapter.

5.1 MODEL FORMULATION

The model formulation of the present problem can be obtained by modifying Equations 4.1 and 4.4 through 4.6, that is,

$$FI_t = \max \delta \qquad (5.1)$$

subject to

$$h_i\big(\mathbf{d}, \mathbf{z}(t), \dot{\mathbf{x}}(t), \mathbf{x}(t), \boldsymbol{\theta}(t)\big) = 0, \quad \mathbf{x}(0) = \mathbf{x}^0, \quad \forall i \in \mathbb{I}; \qquad (5.2)$$

$$\max_{\boldsymbol{\theta}(t)} \ \min_{\mathbf{x}(t),\mathbf{z}(t)} \ \max_{j,t} \ g_j\big(\mathbf{d}, \mathbf{z}(t), \mathbf{x}(t), \boldsymbol{\theta}(t)\big) \le 0; \qquad (5.3)$$

$$\boldsymbol{\theta}^N(t) - \Delta\boldsymbol{\theta}^-(t) \le \boldsymbol{\theta}(t) \le \boldsymbol{\theta}^N(t) + \Delta\boldsymbol{\theta}^+(t); \qquad (5.4)$$

$$-\delta\Delta\boldsymbol{\Theta}^- \le \boldsymbol{\Theta}(H) \le +\delta\Delta\boldsymbol{\Theta}^+ \qquad (5.5)$$

where $\boldsymbol{\Theta}(t) = \int_0^t \big[\boldsymbol{\theta}(\tau) - \boldsymbol{\theta}^N(\tau)\big]d\tau$ and $t \in [0, H]$. Although Equations 5.2 and 5.3 in this model are the same as Equations 4.1 and 4.5, Equations 5.4 and 5.5 are introduced to replace Equation 4.6 for computing FI_t. Note also that a scalar variable δ is used here to adjust the range of the accumulated quantities in Equation 5.5, whereas the transient variations of uncertain parameters are still bounded between the original upper and lower limits (i.e., see Equation 5.4). In computing the dynamic flexibility index FI_d, the former constraints are in fact not considered, whereas the latter are modified with δ instead (i.e., see Equation 4.6).

5.2 NUMERICAL SOLUTION STRATEGIES

The traditional vertex method and the active set method can also be extended to compute the temporal flexibility index (Wu and Chang, 2017). These two alternative approaches are discussed in detail in the following subsections.

5.2.1 Extended Vertex Method—Temporal Version

The vertex locations of a hypercube defined by Equation 5.5 can be expressed mathematically as

$$\Theta(H) = \delta \Delta \Theta^k \qquad (5.6)$$

where $\Delta \Theta^k$ denotes a vector pointing from the origin (i.e., the nominal point) in the $n_p - dimensional$ Euclidean space (where n_p is the number of the uncertain parameters) toward the kth vertex ($k = 1,2,3,\cdots,2^{n_p}$), and each element in $\Delta \Theta^k$ must be the same as the corresponding entry in either $-\Delta \Theta^-$ or $+\Delta \Theta^+$. Furthermore, from the definition of $\Theta(H)$ $\left(= \int_0^H \left[\theta(\tau) - \theta^N(\tau) \right] d\tau \right)$, it is clear that every vertex can be reached with an *infinite* number of time profiles that are bounded by Equation 5.4. Therefore, to be able to implement the temporal version of the vertex method in practical applications, it is, of course, necessary to reduce the search space to a manageable size.

It has been found that, in addition to Equation 5.4, a useful heuristic can be adopted to further constrain the candidate time profiles of uncertain parameters, that is,

$$\theta_n(t) - \theta_n^N(t) = \begin{cases} \Delta \theta_n^k(t), & \text{if } 0 < t_0^n \leq t \leq t_f^n < H \\ 0, & \text{if } 0 \leq t < t_0^n \text{ or } t_f^n < t \leq H \end{cases} \qquad (5.7)$$

where $n = 1,2,\cdots,n_p$ and $k = 1,2,3,\cdots,2^{n_p}$. It should be noted that $\Delta \theta_n^k(t)$ in this equation represents the element n of a vector, which corresponds to vertex k of the hypercube defined by Equation 5.5. More specifically, the position of this vertex can be expressed as

$$\theta(t) = \theta^N(t) + \Delta \theta^k(t) \qquad (5.8)$$

where $\theta(t) = \begin{bmatrix} \theta_1(t) & \theta_2(t) & \cdots & \theta_{n_p}(t) \end{bmatrix}^T$,

$\theta^N(t) = \begin{bmatrix} \theta_1^N(t) & \theta_2^N(t) & \cdots & \theta_{n_p}^N(t) \end{bmatrix}^T$,

$\Delta \theta^k(t) = \begin{bmatrix} \Delta \theta_1^k(t) & \Delta \theta_2^k(t) & \cdots & \Delta \theta_{n_p}^k(t) \end{bmatrix}^T$ and $t_0^n \leq t \leq t_f^n$ $(n = 1,2,\cdots,n_p)$.

As mentioned previously in Chapter 4, $\Delta \theta^k(t)$ can be viewed as a vector in the *functional space* of $\theta(t)$, which starts from the nominal point $\theta^N(t)$ and ends at the kth vertex, and each element of $\Delta \theta^k(t)$ should be selected from the corresponding entry in either $-\Delta \theta^-(t)$ or $+\Delta \theta^+(t)$. Notice also that as clearly indicated in Equation 5.7, the allowed deviation in each uncertain parameter may begin and terminate at instances that are not the same as those of the other parameters. Finally, although the justification for the heuristics earlier is derived from an intuitive belief, that is, the most severe disturbance a realistic process can withstand is usually the one with the largest possible magnitude within the shortest period, its validity has been verified in extensive case studies.

In principle, FI_t can be determined by solving Equations 5.1 through 5.3 and Equations 5.6 through 5.8 via discretization, and for illustration simplicity, let us again utilize the trapezoidal rule for this purpose here. Because the starting and ending times of parameter deviations, that is, t_0^n, t_f^n and $n = 1,2,\cdots,n_p$, are not given *a priori*, an extra binary variable $\varepsilon_p^n \in \{0,1\}$ must be introduced at every discretized time t_p^n to reflect if the corresponding deviation, that is, $\Delta\theta_n^k(t_p^n)$, takes place. Specifically,

$$\theta_n(t_p^n) - \theta_n^N(t_p^n) = \begin{cases} \Delta\theta_n^k(t_p^n) & \text{if } \varepsilon_p^n = 1 \\ 0 & \text{if } \varepsilon_p^n = 0 \end{cases} \tag{5.9}$$

With these binary variables, additional logic constraints can be incorporated in a mathematical programming model to enforce the heuristic in Equation 5.7. If the disturbance in θ_n earlier starts at a particular discretized time $t_{p'}^n$, then Equations 5.7 and 5.8 can be expressed as follows:

$$\varepsilon_{p'}^n = 1 \tag{5.10}$$

$$\varepsilon_0^n = \varepsilon_1^n = \ldots = \varepsilon_{p'-1}^n = 0 \tag{5.11}$$

$$(1 - \varepsilon_p^n) + \varepsilon_{p+1}^n \leq 1 \tag{5.12}$$

where $p' \in \{1,2,\cdots,M-1\}$ and $p = p', p'+1,\cdots,M-1$. Equation 5.12 clearly implies the following:

1. If deviation is not present at t_p^n (i.e., $\varepsilon_p^n = 0$), then there will not be any at the next instance t_{p+1}^n (i.e., $\varepsilon_{p+1}^n = 0$).
2. If otherwise (i.e., $\varepsilon_p^n = 1$), then the disturbance at t_{p+1}^n may or may not take place (i.e., $\varepsilon_{p+1}^n = 0$ or 1).

Consequently, Equation 5.7 can be rewritten as

$$\theta_n(t_p^n) = \theta_n^N(t_p^n) + \varepsilon_p^n \Delta\theta_n^k(t_p^n) \tag{5.13}$$

where $p = 1,2,\cdots,M$, $n = 1,2,\cdots,n_p$, and $k = 1,2,3,\cdots,2^{n_p}$. The accumulated parameter variations can then be expressed accordingly as

$$\Theta_n(H) = \sum_{p=1}^{M} \varepsilon_p^n \Delta\theta_n^k(t_p^n) \tag{5.14}$$

where $n = 1, 2, \cdots, n_p$. Finally, the discretized version of Equation 5.5 should be

$$-\delta\Delta\Theta_n^- \leq \sum_{p=1}^{M} \varepsilon_p^n \Delta\theta_n^k\left(t_p^n\right) \leq \delta\Delta\Theta_n^+ \tag{5.15}$$

where $n = 1, 2, \cdots, n_p$.

By making use of these formulations, a mixed integer nonlinear programming (MINLP) model can be constructed to implement the temporal version of the vertex method, that is,

$$FI_t = \min_{k, \varepsilon} \max_{\mathbf{Z}, \mathbf{X}, \mathbf{E}} \delta \tag{5.16}$$

subject to Equations 4.14 through 4.16 and Equations 5.10 through 5.15. More specifically,

$$\mathbf{E} = \begin{bmatrix} \varepsilon_1^1 & \varepsilon_2^1 & \cdots & \varepsilon_M^1 \\ \varepsilon_1^2 & \varepsilon_2^2 & \cdots & \varepsilon_M^2 \\ \vdots & \vdots & \cdots & \vdots \\ \varepsilon_1^{n_p} & \varepsilon_2^{n_p} & \cdots & \varepsilon_M^{n_p} \end{bmatrix};$$

$$\mathbf{X} = \begin{bmatrix} \mathbf{x}(t_1) & \mathbf{x}(t_2) & \cdots & \mathbf{x}(t_M) \end{bmatrix};$$

$$\mathbf{Z} = \begin{bmatrix} \mathbf{z}(t_1) & \mathbf{z}(t_2) & \cdots & \mathbf{z}(t_M) \end{bmatrix};$$

$$\varepsilon = \begin{bmatrix} \varepsilon_{p_1'}^1 & \varepsilon_{p_2'}^2 & \cdots & \varepsilon_{p_{n_p}'}^{n_p} \end{bmatrix}^T$$

where $p_1', p_2', \cdots, p_{n_p}' \in \{1, 2, \cdots, M-1\}$; $k = 1, 2, 3, \cdots, 2^{n_p}$.

Notice that Equation 5.16 is a two-level optimization problem. If there are relatively few uncertain parameters, the minimization calculations in the upper level may be accomplished with a simple-minded exhaustive enumeration procedure, and the lower-level maximization can be performed by using a commercial solver. This approach is summarized in Figure 5.1.

5.2.2 EXTENDED ACTIVE SET METHOD—TEMPORAL VERSION

Because the temporal version of the feasibility functional can be formulated with Equations 4.21 through 4.24 as well, the resulting Karush-Kuhn-Tucker (KKT) conditions should be the same as those described in Section 4.2.3. As a result, the

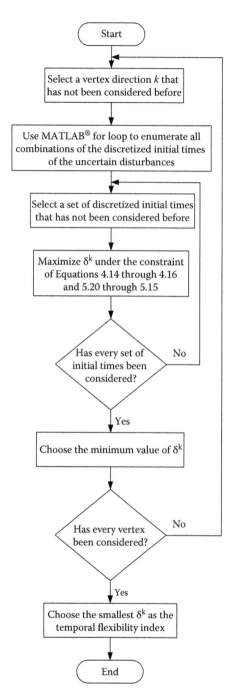

FIGURE 5.1 Temporal version of the extended vertex method. (From Kuo, Y.C., 2015. Applications of the dynamic and temporal flexibility indices, Department of Chemical Engineering. National Cheng Kung University, Tainan.)

optimization problem to be solved in the temporal version of the active set approach (Kuo and Chang, 2016) can be expressed as follows:

$$FI_t = \min_{\delta,\mu_u(t),\mu_i(t),\lambda_j(t),s_j(t),y_j(t),x_i(t),z_c(t),\theta_n(t)} \delta \tag{5.17}$$

$$\frac{dx_i(t)}{dt} = \varphi_i\left(\mathbf{d},\mathbf{x}(t),\mathbf{z}(t),\theta(t)\right), \quad x_i(0) = x_i^0, \quad \forall i \in \mathbb{I}; \tag{5.18}$$

$$g_j\left(\mathbf{d},\mathbf{z}(t),\mathbf{x}(t),\theta(t)\right) + s_j(t) = 0, \quad s_j(t) \geq 0, \quad \forall j \in \mathbb{J}; \tag{5.19}$$

$$\dot{\mu}_u(t) = \sum_{j' \in \mathbb{J}} \lambda_{j'}(t), \quad \mu_u(0) = 0, \quad \mu_u(H) = 1; \tag{5.20}$$

$$\dot{\mu}_i(t) = -\sum_{i' \in \mathbb{I}} \mu_{i'}(t)\frac{\partial \varphi_{i'}}{\partial x_i}(t) - \sum_{j' \in \mathbb{J}} \lambda_{j'}(t)\frac{\partial g_{j'}}{\partial x_i}(t), \quad \mu_i(H) = 0, \quad \forall i \in \mathbb{I}; \tag{5.21}$$

$$\sum_{i' \in \mathbb{I}} \mu_{i'}(t)\frac{\partial \varphi_{i'}}{\partial z_c}(t) + \sum_{j' \in \mathbb{J}} \lambda_{j'}(t)\frac{\partial g_{j'}}{\partial z_c}(t) = 0, \quad c = 1,2,\cdots,n_z; \tag{5.22}$$

$$s_j(t) - U\left(1 - y_j(t)\right) \leq 0, \quad \lambda_j(t) - y_j(t) \leq 0, \quad y_j(t) \in \{0,1\}, \quad \lambda_j(t) \geq 0, \quad \forall j \in \mathbb{J}; \tag{5.23}$$

$$\theta_n^N(t) - \Delta\theta_n^-(t) \leq \theta_n(t) \leq \theta_n^N(t) + \Delta\theta_n^+(t), \quad n = 1,2,\cdots,n_p; \tag{5.24}$$

$$-\delta\Delta\Theta_n^- \leq \int_0^H \left[\theta_n(\tau) - \theta_n^N(\tau)\right] d\tau \leq \delta\Delta\Theta_n^+ \tag{5.25}$$

where $0 < t \leq H$. Although Equations 5.18 through 5.23 are identical to Equations 4.27 through 4.32, these constraints are still presented here for the sake of completeness. Note that to compute FI_t, an implementable formulation should again be generated by discretizing all of the earlier constraints with the trapezoidal rule.

5.3   AN ILLUSTRATIVE EXAMPLE

Let us again consider the buffer operations described previously in Section 4.3 (see Figure 4.2). The system description is partially repeated here for illustration clarity.
    In particular, the dynamic model of this buffer system can be written as

$$A\frac{dh}{dt} = \theta(t) - k\sqrt{h} \tag{5.26}$$

where h denotes the height of liquid level (m); A $(= 5 \text{ m}^2)$ is the cross-sectional area of the tank; k $(= \sqrt{5}/10 \text{ m}^{5/2} \text{ min}^{-1})$ is a proportionality constant; and θ denotes the feed flow rate (m^3 min^{-1}) and it is treated as the only uncertain parameter in the present example. The allowed range of the state variable—that is, the height of the liquid level—and the time horizon are the same as those given in Section 4.3, that is,

1. The height of tank itself is 10 m $(h \leq 10)$.
2. Due to the operational requirement of downstream unit(s), the outlet flow rate of the buffer tank must be kept above $\sqrt{5}/10 \text{ m}^3\text{min}^{-1}$. Thus, the minimum allowable height of its liquid level should be 1 m $(h \geq 1)$.
3. The time horizon covers a period of 800 minutes $(0 \leq t \leq 800)$.

Case 5.1: Continuous Operation

Let us again assume that in the continuous operation under consideration, the nominal steady-state value of the feed rate is $\theta^N(t) = 0.5 \text{ m}^3 \text{ min}^{-1}$ and the corresponding anticipated positive and negative deviations are set at $\Delta\theta^+(t) = \Delta\theta^-(t) = 0.5 \text{ m}^3\text{min}^{-1}$, respectively. Therefore, the range of the uncertain parameter is

$$0 \leq \theta(t) \leq 1 \qquad (5.27)$$

Moreover, the nominal height of the liquid level at the steady state should be 5 m. To facilitate computation of the temporal flexibility index in the present case, let us further set the accumulated positive and negative deviations in liquid volumes (m^3) to be

$$\Delta\Theta^+ = \Delta\Theta^- = 62.5 \qquad (5.28)$$

In other words, the feed rate in the anticipated worst-case scenario is required to be reduced to the lower limit (0.0 m^{3}min^{-1}) or raised to the upper bound (1.0 m^{3}min^{-1}) for a period of 125 $\left(= \dfrac{62.5}{0.5} \right)$ minutes. By executing the GAMS code in Appendix 5.1, it can be found that $FI_t = 0.444$, which implies that the given system can only withstand the most severe disturbance for a shorter period of 55.5 $(= 125 \times 0.444)$ minutes. Figure 5.2 shows the simulation results of two corresponding scenarios. One can observe that (1) the water level just touches 1 m (i.e., the lower bound of h) at 55.5 min after introducing the largest negative disturbance at 0 min, and (2) the water level always stays considerably below the upper limit (i.e., $h < 10$) if the largest positive disturbance lasts only 55.5 minutes. The former is therefore the worst-case scenario.

As mentioned previously in Case 4.1, the value of FI_d (which is 0.415 for the base-case design) can be improved to the implied target of 1 by increasing the cross-sectional area of the buffer vessel from 5 m^2 to 61 m^2. Because a relatively large tank is called for, the required capital investment may not be justifiable. However, if it can be predicted on the basis of operation experience that the largest disturbances rarely last for the entire horizon, then the design target of $FI_t = 1$ may be acceptable. In this case, the corresponding area should be 11.3 m², and the required cost is obviously much lower. Figure 5.3 shows the simulation results of the worst-case scenarios for $FI_t = 1$ in Case 5.1.

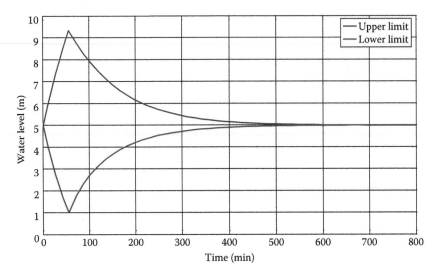

FIGURE 5.2 Simulation results of the worst-case scenarios in Case 5.1 ($FI_t = 0.444$, $A = 5.0$ m², $t_0 = 0$ min, $t_f = 55.5$ min). (Reprinted with permission from Kuo and Chang, 2016, 670–682. Copyright 2016 American Chemical Society.)

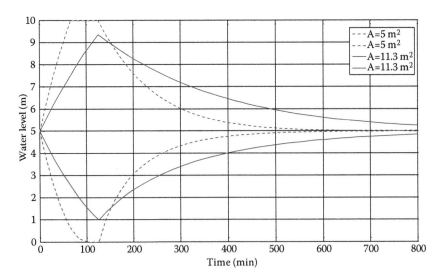

FIGURE 5.3 Simulation results of the worst-case scenarios in Case 5.1 ($FI_t = 1$, $A = 5.0$ or 11.3 m², $t_0 = 0$ min, $t_f = 55.5$ min). (Reprinted with permission from Kuo and Chang, 2016, 670–682. Copyright 2016 American Chemical Society.)

Case 5.2: Periodic Operation

The time profiles of the nominal feed rate and its anticipated positive and negative deviations in one operation cycle have already been reported in Chapter 4; for illustration convenience, they are repeated in Figure 5.4. To facilitate the concrete computation of the temporal flexibility index in the present case, let us assign the accumulated positive and negative deviations in liquid volumes (m^3) to be

$$\Delta\Theta^+ = \Delta\Theta^- = 20.0 \tag{5.29}$$

By discretizing Equation 5.26 according to the trapezoidal rule (see Section 4.2.1.2) and then implementing the extended vertex method (see Subsection 5.2.1), one can find that the corresponding temporal flexibility index is 0.185 for which the disturbance only exists in the time interval between 562 and 599 minutes (see the GAMS and MATLAB codes in Appendix 5.2). These results imply that when the largest deviation in the feed rate is present in the period noted earlier, an accumulated volume decrease of 3.7 ($= 20 \times 0.185$) m^3 should cause the water level to reach the lower limit of 1 m at the end point of this time interval. This prediction can be clearly observed in the simulation results presented in Figure 5.5. Finally, notice that the temporal flexibility in this case can also be enhanced with a larger tank. From the optimum solution of the proposed programming model, one can deduce that at least a cross-sectional area of 6.98 m^2 should be adopted to achieve the designated design target when $FI_t = 1$ (see Figure 5.6). Finally, it should be noted that the same FI_t (0.185) can be obtained more efficiently in this case with the extended active set method. The needed computation time in GAMS 23.9.5 on an i7-4770 PC was found to be 265 sec, which is much shorter than that consumed with the extended vertex method (1920 sec).

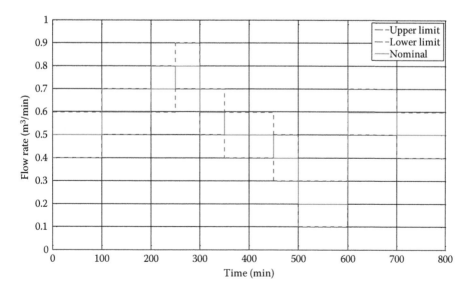

FIGURE 5.4 Nominal feed rate and its upper and lower limits in a single period in Case 5.2. (Reprinted with permission from Kuo and Chang, 2016, 670–682. Copyright 2016 American Chemical Society.)

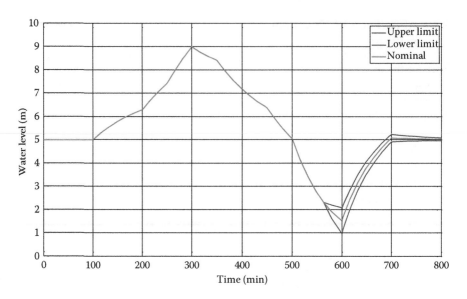

FIGURE 5.5 Simulation results of the worst-case scenarios in Case 5.2 ($FI_t = 0.185$, $A = 5.0$ m^2, $t_0 = 562$ min, $t_f = 599$ min). (Reprinted with permission from Kuo and Chang, 2016, 670–682. Copyright 2016 American Chemical Society.)

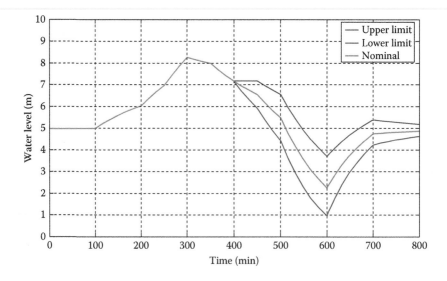

FIGURE 5.6 Simulation results of the worst-case scenarios in Case 5.2 ($FI_t = 1$), $A = 6.95$ m^2, $t_0 = 399$ min, $t_f = 599$ min). (Reprinted with permission from Kuo and Chang, 2016, 670–682. Copyright 2016 American Chemical Society.)

5.4 APPLICATIONS OF DYNAMIC AND TEMPORAL FLEXIBILITY ANALYSES IN PROCESS DESIGNS

Because the dynamic and temporal indices are defined to characterize distinct features in the transient behaviors of unsteady systems, their numerical values should be interpreted differently. The following conclusions can be drawn from the earlier discussions:

- A less-than-1 dynamic flexibility index ($FI_d < 1$) implies that the given system cannot withstand at least a set of disturbances in the worst-case scenario constrained by Equation 4.3. Although the corresponding value of the temporal flexibility index is unbounded—that is, its value may (or may not) be smaller than one—the larger of the two indices can only be determined on a case-by-case basis.
- A less-than-1 temporal flexibility index ($FI_t < 1$) implies that the given system cannot withstand at least a set of the disturbances in the worst-case scenario constrained by Equations 4.3, 4.39, and 4.40. Although the corresponding value of the dynamic flexibility index should also be less than 1 ($FI_d < 1$), the larger of the two can only be determined on a case-by-case basis.
- $FI_t = 1$ implies $FI_d \leq 1$, whereas $FI_d = 1$ implies $FI_t \geq 1$.
- Due to the requirements imposed in Equations 4.41 and 4.42, a *fixed* larger-than-1 upper limit of FI_t should be reached for all $FI_d \geq 1$.

Based on the earlier observations, a generic design procedure can be accordingly summarized as follows:

For any given design, the dynamic flexibility index should be computed first. If it can be determined that $FI_d \geq 1$, then, of course, no changes are needed. If otherwise and the corresponding revamp measures for achieving $FI_d = 1$ are expensive, then the temporal flexibility index should be determined according to the proposed mathematical program. The proper design modifications for realizing $FI_t = 1$ can then be identified on the basis of their economic implications.

APPENDIX 5.1: GAMS CODE USED IN CASE 5.1 OF THE BUFFER TANK EXAMPLE

```
sets
m discrete time /1*800/
;
parameters
Ar area of tank /5/
k;
k=5**0.5/10;
positive variables
fl  flexibility index
h(m)  tank level
```

```
the(m)   flow rate
;
variables
z ;
h.l(m)=5;
binary variables
e(m)
;
equations
eqcon(m),ineqcon1(m),ineqcon2(m)
lo(m),el(m),se,h0,e0,min
;
eqcon(m)$(ord(m)<800)..Ar*(h(m+1)-h(m))=e=0.5*(the(m+1)+the
(m))-0.5*k*(h(m+1)**0.5+h(m)**0.5);
ineqcon1(m)..(h(m)-10)=l=0;
ineqcon2(m)..(1-h(m))=l=0;
lo(m)..the(m)=e=0.5-0.5*e(m);
el(m)$(ord(m)<800)..(1-e(m))+e(m+1)=l=1;
se..fl=e=sum(m,0.5*e(m))/62.5;
h0..h('1')=e=5;
e0..e('1')=e=1;
min..z=e=fl;
model aaa /all/;
solve aaa using minlp maximizing z;
display z.l,h.l,the.l,e.l;
```

APPENDIX 5.2:   GAMS AND MATLAB CODES USED IN CASE 5.2 OF THE BUFFER TANK EXAMPLE

GAMS CODE

```
sets
m discrete time /1*800/
;
parameters
Ar area of tank /6.95/
s initial time /1/
dt uncertain deviation /-0.1/
k;
k=5**0.5/10;
positive variables
fl  flexibility index
h(m)   tank level
the(m)   flow rate
;
variables
z ;
h.l(m)=5;
binary variables
```

```
e(m)
;
equations
eqcon(m),ineqcon1(m),ineqcon2(m)
up1(m),up2(m),up3(m),up4(m),up5(m),up6(m),up7(m),up8(m),up9(m),
up10(m)
el(m),h0,ee(m),e0,se1,min
;
eqcon(m)$(ord(m)<800)..Ar*(h(m+1)-h(m))=e=0.5*(the(m+1)+
the(m))-0.5*k*(h(m+1)**0.5+h(m)**0.5);
ineqcon1(m)..(h(m)-10)=l=0;
ineqcon2(m)..(1-h(m))=l=0;
up1(m)$(ord(m)<101)..the(m)=e=0.5+dt*e(m);
up2(m)$(ord(m)<201 and ord(m)>100)..the(m)=e=0.6+dt*e(m);
up3(m)$(ord(m)<251 and ord(m)>200)..the(m)=e=0.7+dt*e(m);
up4(m)$(ord(m)<301 and ord(m)>250)..the(m)=e=0.8+dt*e(m);
up5(m)$(ord(m)<351 and ord(m)>300)..the(m)=e=0.6+dt*e(m);
up6(m)$(ord(m)<451 and ord(m)>350)..the(m)=e=0.5+dt*e(m);
up7(m)$(ord(m)<501 and ord(m)>450)..the(m)=e=0.4+dt*e(m);
up8(m)$(ord(m)<601 and ord(m)>500)..the(m)=e=0.2+dt*e(m);
up9(m)$(ord(m)<701 and ord(m)>600)..the(m)=e=0.6+dt*e(m);
up10(m)$(ord(m)>700)..the(m)=e=0.5+dt*e(m);
el(m)$(ord(m)>s-1 and ord(m)<800)..(1-e(m))+e(m+1)=l=1;
h0..h('1')=e=5;
ee(m)$(ord(m)<s)..e(m)=e=0;
e0(m)$(ord(m)=s)..e(m)=e=1;
se1..sum(m,-dt*e(m))=e=20*fl;
min..z=e=fl;
model aaa /all/;
solve aaa using minlp maximizing z;
scalars modelStat, solveStat;
modelStat = aaa.modelstat;
solveStat = aaa.solvestat;
display z.l,h.l,the.l,e.l,modelStat, solveStat;
```

MATLAB CODE

```
for k=1:800
    ns.name = 's';
    ns.type = 'parameter';
    ns.val = k;
    ns.form = 'full';
    ns.dim = 0;

    wgdx('voldata',ns);

    gams_output = 'std';
    gams('newtank');
    solGDX='volsol.gdx';
```

```
      rs = struct('name','z','field','l','form','full');
      r = rgdx (solGDX, rs);
      obj = r.val;
      result(k)=obj;
end
```

REFERENCES

Adi, V.S.K., Chang, C.T., 2013. A mathematical programming formulation for temporal flex-
    ibility analysis. *Computers & Chemical Engineering* 57, 151–158.
Kuo, Y.C., 2015. Applications of the dynamic and temporal flexibility indices, Department of
    Chemical Engineering. National Cheng Kung University, Tainan.
Kuo, Y.C., Chang, C.T., 2016. On heuristic computation and application of flexibility indices
    for unsteady process design. *Industrial & Engineering Chemistry Research* 55, 670–682.
Wu, R.S., Chang, C.T., 2017. Development of mathematical programs for evaluating dynamic
    and temporal flexibility indices based on KKT conditions. *Journal of the Taiwan
    Institute of Chemical Engineers* 73, 86–92.

6 Systematic Revamp Strategies for Improving Operational Flexibility of Existing Water Networks

6.1 BACKGROUND

Water is an essential natural resource needed in almost every processing plant. For instance, it is used for crude oil desalting in petroleum refineries; for liquid–liquid extraction in hydrometallurgical processes; for cooling, quenching, and scrubbing in the iron and steel industries; and for various washing operations in food processing facilities. Due to the bleak forecast concerning the water crisis in the near future and the increasingly stringent environmental regulations for wastewater disposal, efficient water utilization is, of course, an indispensable criterion that must be adopted in designing any industrial process (Byers et al., 2003).

Various industrial water management issues have already been addressed extensively in the literature. In particular, some mathematical programming models were developed to optimally route the process streams in a water network for the purpose of minimizing the freshwater consumption rate and/or wastewater generation rate. One of the pioneering papers in this area was published by Takama et al. (1980), who studied the optimal water allocation problem in a petroleum refinery. Wang and Smith (1995) suggested considering water reuse, regeneration–reuse, and regeneration–recycling in water network designs as viable wastewater minimization strategies. They also proposed a heuristic methodology for designing effluent treatment systems in which wastewater was processed in a distributed manner. Alva-Argaez et al. (1998) used a mathematical programming approach to optimize a superstructure in which all possibilities of water treatment and reuse were embedded. Bagajewicz et al. (1999) developed a systematic method to transform the nonlinear model for multicontaminant large-scale water system designs to a linear program (LP). Another important work was carried out by Huang et al. (1999), who presented a comprehensive programming approach to synthesize the optimal water usage and treatment networks in chemical processes. Feng and Seider (2001) assessed the feasibility of simplifying network configurations in large plants with internal water mains. Karuppiah and Grossmann (2006) studied the optimal synthesis problem of integrated water systems, where water using and treatment operations were incorporated into a single network in such a way that the total annual cost could be minimized.

From these studies, it can be observed that complex configurations are often needed in the optimal water networks to facilitate effective reuse–recycle and reuse–regeneration. These elaborate process structures obviously hamper resilient operation and control under uncertain disturbances from the environment. In the past, very few studies have been performed to deal with this important issue. Tan and Cruz (2004) formulated two different versions of the symmetry fuzzy linear programming (SFLP) models for the purpose of synthesizing robust water-reuse networks based on imprecise process data. Al-Redhwan et al. (2005) developed a three-step procedure to design water networks under uncertain operating temperatures and pressures. Tan et al. (2007) used the Monte Carlo simulation techniques to analyze the vulnerability of water networks with noisy mass loads. Karuppiah and Grossmann (2008) proposed a spatial branch-and-cut algorithm to locate the global optimum. Zhang et al. (2009) suggested using the concepts of maximum tolerance amount of a water unit (MToAWU), rank of unit (RU), and outflow branch number of a unit (OBNU) to quantify the resilience of a given water network.

To address these issues, Chang et al. (2009) developed a generalized mixed-integer nonlinear programming model (MINLP) for assessing and improving the operational flexibility of water network designs. They found that any given network can be enhanced with two revamp strategies: (1) relaxation of the upper limit of the freshwater supply rate and (2) installation of auxiliary pipelines and/or elimination of existing ones. By making use of the insights gained from active constraints, Riyanto and Chang (2010) later developed a heuristic manual strategy in a subsequent study to raise the steady-state flexibility indices (FI_ss) of existing single-contaminant water networks. In addition, Li and Chang (2011) constructed a new nonlinear programming formulation model by incorporating process knowledge into the conventional vertex method to simplify FI_s calculation.

Although satisfactory results have been reported in the works mentioned earlier, it should be noted that only the single-contaminant systems were considered and, more importantly, the total number of candidate configurations may be too large to be evaluated in a manual evolution procedure. Also, if the active set method is to be utilized for calculation of the steady-state flexibility index, the Karush-Kuhn-Tucker (KKT) conditions must be invoked to construct a tedious model for the multicontaminant case. Thus, it is evident that the iterative solution processes of the previously mentioned MINLP model for computing FI_s may not always converge. To overcome these difficulties, an alternative computation strategy has been devised by Jiang and Chang (2013) by solving an NLP model iteratively. On the basis of this modified computation method, the proper revamp options can be identified evolutionarily with a genetic algorithm.

The aforementioned revamp strategies for the single- and multicontaminant water networks are presented in detail in the next sections.

6.2 AUGMENTED SUPERSTRUCTURE

Because it is tedious and inefficient to construct different versions of the flexibility index model for various candidate designs then carry out the needed optimization runs, a generalized model should be formulated and used as a design tool for all possible structures under consideration. It is necessary first to build an augmented superstructure in which all possible *new* connections are embedded to develop such a model by an existing network.

6.2.1 LABEL SETS

For illustration convenience, let us first define the following label sets:

$$W_1 = \left\{ w_1 \middle| w_1 \text{ is the label of an existing primary water source} \right\}$$

$$W_2 = \left\{ w_2 \middle| w_2 \text{ is the label of an existing secondary water source} \right\}$$

$$S = \left\{ s \middle| s \text{ is the label of an existing sink} \right\}$$

$$U = \left\{ u \middle| u \text{ is the label of an existing water } using \text{ unit} \right\}$$

$$T = \left\{ t \middle| t \text{ is the label of an existing treatment unit} \right\}$$

$$X = \left\{ x \middle| x \text{ is the label of an added treatment unit} \right\}$$

Based on these definitions, one can then assemble the following sets for characterizing the superstructure:

- The label set of all water sources embedded in the superstructure, that is,

$$W = W_1 \cup W_2$$

- The label set of all processing units embedded in the superstructure, that is,

$$P = U \cup T \cup X$$

- The label set of all existing processing units, that is,

$$P' = U \cup T$$

- The label set of all split nodes in the superstructure, that is,

$$M = W \cup U \cup T \cup X$$

- The label set of all split nodes at the outlets of existing units, that is,

$$M' = W \cup U \cup T$$

- The label set of all mixing nodes in the superstructure, that is,

$$N = U \cup T \cup X \cup S$$

- The label set of all mixing nodes at the inlets of existing units, that is,

$$N' = U \cup T \cup S$$

Finally, if multiple contaminants are present in a water network, then an additional label set should be considered:

$$K = \left\{ k \middle| k \text{ is the label of a water contaminant} \right\}$$

6.2.2 Superstructure Construction

A conventional superstructure for grassroots designs can be constructed according to the following steps:

> Step 1: Connect the split node at the outlet of every primary water source in W_1 to the mixing node at the inlet of every processing unit in P.
> Step 2: Connect the split node at the outlet of every secondary water source in W_2 to every mixing node in N.
> Step 3: Connect the split node at the outlet of every processing unit in P to every mixing node in N.

This conventional configuration can then be transformed into an augmented superstructure by classifying all embedded connections into three types. More specifically, these connection types can be associated, respectively, with (1) the existing pipelines; (2) the new pipelines between existing units—that is, the connections from M' to N', which are *not* present in the given network—and (3) the new pipelines between the added treatment units and other units, that is, the connections from M to X and from X to N. Finally, the notational convention in Table 6.1 is followed throughout this chapter to facilitate unambiguous formulation of the corresponding mathematical model, and a simple example is given next for further clarification.

Example 6.1

Let us consider the existing water network presented in Figure 6.1, in which a freshwater source (W_1), a water-using unit (U_1), a wastewater treatment unit (T_1), and a sink (S_1) are involved. The corresponding augmented superstructure can be found in Figure 6.2. The symbols FT_{W_1}, FT_{U_1}, FT_{T_1}, and FT_{S_1} in this superstructure, respectively, denote the throughputs in W_1, U_1, T_1, and S_1, whereas F_{W_1,U_1}, F_{U_1,S_1}, and F_{T_1,S_1} denote the flow rates in the *existing* pipelines (Type-1 connections), that is, from W_1 to U_1, from U_1 to T_1, and from T_1 to S_1, respectively. Based on the classification criteria mentioned earlier, there can be four new pipelines connecting the split nodes in M' and the mixing nodes in N'. Specifically, these Type-2 connections are (W_1,T_1), (U_1, S_1), (T_1,U_1), and (T_1,T_1); moreover, the corresponding flow rates are expressed as f_{W_1,T_1}, f_{U_1,S_1}, f_{T_1,U_1}, and f_{T_1,T_1}, respectively. If an extra treatment unit X_1 is allowed to be installed in this network, then at most six new pipelines may be added, that

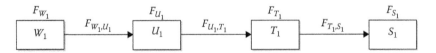

FIGURE 6.1 Existing water network in Example 6.1. (Reprinted from *Chemical Engineering Science*, Vol. 102, Jiang, D., Chang, C.T., An algorithmic revamp strategy for improving operational flexibility of multi-contaminant water networks, 289–299. Copyright 2013 with permission from Elsevier.)

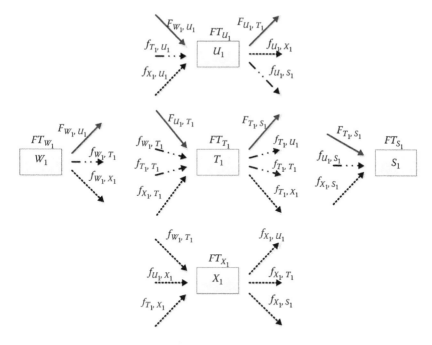

FIGURE 6.2 Augmented superstructure for Example 6.1. (Reprinted from *Chemical Engineering Science*, Vol. 102, Jiang, D., Chang, C.T., An algorithmic revamp strategy for improving operational flexibility of multi-contaminant water networks, 289–299. Copyright 2013 with permission from Elsevier.)

is, (W_1,X_1), (U_1,X_1), (T_1,X_1), (X_1,U_1), (X_1,T_1), and (X_1,S_1); moreover, they should be regarded as Type-3 connections. Their flow rates are denoted, respectively, as f_{W_1,X_1} f_{U_1,X_1} f_{T_1,X_1} f_{X_1,U_1} f_{X_1,T_1} and f_{X_1,S_1}.

6.3 MODEL CONSTRAINTS

The equality and inequality constraints of the mathematical programming model can be formulated by the augmented superstructure and the notational convention given in Table 6.1. A brief summary of these constraints can be found in the sequel.

6.3.1 BINARY DESIGN PARAMETERS

In the proposed computation procedure, the flexibility index of a revamped network is calculated by the existing network and a collection of new pipelines and/or new treatment units, which are chosen from the outset. To facilitate model formulation, let us introduce the following binary *design parameters*:

$$d_{m,n} = \begin{cases} 1 & \text{if the } new \text{ connection between } m \in M \text{ and } n \in N \text{ is chosen} \\ 0 & \text{otherwise} \end{cases} \quad (6.1)$$

TABLE 6.1

Notational Convention

	Notation	Definition
Subscripts	k	The label corresponding to a water contaminant
	m	The label corresponding to a split node in the superstructure
	n	The label corresponding to a mixing node in the superstructure
	p	The label corresponding to a processing unit
	s	The label corresponding to an existing sink
	t	The label corresponding to an existing treatment unit
	w	The label corresponding to an existing water source
	x	The label corresponding to an added treatment unit
Continuous variables or parameters	C	The contaminant concentration
	f	The water flow rate of a new connection
	ft	The water throughput in a new unit
	F	The water flow rate of an existing connection
	FT	The total water throughput in an existing unit
	ML	The mass load in a water-using unit
	R	The removal ratio of a contaminant in a treatment unit
	θ	The uncertain multiplier
Binary parameters	d	The existence/nonexistence of a new connection (pipeline)
Superscripts	in	The inlet of a unit
	out	The outlet of a unit
	max	Upper bound
Sets	K	The label set for all contaminants
	M	The label set for all split nodes in the superstructure
	N	The label set for all mixing nodes in the superstructure
	P	The label set for all processing units
	S	The label set for all existing sinks
	T	The label set for all existing treatment units
	W	The label set for all existing water sources
	X	The label set for all available new treatment units

Source: Reprinted from *Chemical Engineering Science*, Vol. 102, Jiang, D., Chang, C.T., An algorithmic revamp strategy for improving operational flexibility of multi-contaminant water networks, 289–299. Copyright 2013 with permission from Elsevier.

Note that these parameters should be *fixed* before computing the flexibility index. Also, the following flow constraints should be imposed in the flexibility index model:

$$f^L d_{m,n} \leq f_{m,n} \leq f^U d_{m,n} \tag{6.2}$$

6.3.2 PRIMARY SOURCES

The freshwater supplies secured by a chemical plant are regarded as the primary water sources in the model. It is also assumed that any effluent is not allowed to be mixed with freshwater to meet the discharge limit required by environmental regulation. The mass balance at the outlet split node of every primary source can be written as:

$$FT_{w_1} = \sum_{p \in P'} F_{w_1,p} + \sum_{n \in P} f_{w_1,n} \tag{6.3}$$

In practical applications, an upper bound should be imposed upon the freshwater supply rate:

$$FT_{w_1} \leq FT_{w_1}^{\max} \tag{6.4}$$

where $w_1 \in W_1$.

6.3.3 SECONDARY SOURCES

The pollutant concentrations in secondary water are usually higher than those in the primary source. The mass balance at the outlet split node can be expressed as

$$FT_{w_2} = \sum_{p \in P'} F_{w_2,p} + \sum_{s \in S} F_{w_2,s} + \sum_{n \in N} f_{w_2,n} \tag{6.5}$$

where $w_2 \in W_2$.

6.3.4 SINKS

The wastewater can be discharged into the environment or other effluent treatment facilities. The mass balance constraints at the inlet mixing node of each sink can be expressed as

$$FT_s = \sum_{p \in P'} F_{p,s} + \sum_{w_2 \in W_2} F_{w_2,s} + \sum_{m \in M} f_{m,s} \tag{6.6}$$

$$FT_s C_{s,k} = \sum_{p \in P'} F_{p,s} C_{p,k} + \sum_{w_2 \in W_2} F_{w_2,s} C_{w_2,k} + \sum_{m \in M} f_{m,s} C_{m,k} \tag{6.7}$$

where $s \in S$ and $k \in K$. Obviously, an upper bound should be imposed on every contaminant concentration at the sink to conform to the environmental regulations:

$$C_{s,k} \leq C_{s,k}^{\max} \tag{6.8}$$

6.3.5 WATER-USING UNITS

The mass balances for characterizing the water-using units are given as follows:

$$FT_u \left(C_{u,k}^{\text{out}} - C_{u,k}^{\text{in}} \right) = ML_{u,k} \tag{6.9}$$

$$FT_u = \sum_{\substack{w \in W}} F_{w,u} + \sum_{\substack{p \in P' \\ p \neq u}} F_{p,u} + \sum_{\substack{m \in M \\ m \neq u}} f_{m,u}$$

$$= \sum_{\substack{p \in P' \\ p \neq u}} F_{u,p} + \sum_{\substack{s \in S}} F_{u,s} + \sum_{\substack{n \in N \\ n \neq u}} f_{u,n} \tag{6.10}$$

where $u \in U$ and $k \in K$. The upper limits of $C_{u,k}^{\text{in}}$ and $C_{u,k}^{\text{out}}$ must also be imposed, that is,

$$C_{u,k}^{\text{in}} \leq C_{u,k}^{\text{in,max}} \tag{6.11}$$

$$C_{u,k}^{\text{out}} \leq C_{u,k}^{\text{out,max}} \tag{6.12}$$

6.3.6 WATER TREATMENT UNITS

The following mass balance constraints are adopted in this work to model the water treatment units:

$$C_{t,k}^{\text{in}} \left(1 - R_{t,k} \right) = C_{t,k}^{\text{out}} \tag{6.13}$$

$$FT_t = \sum_{\substack{w \in W}} F_{w,t} + \sum_{\substack{p \in P'}} F_{p,t} + \sum_{\substack{m \in M}} f_{m,t}$$

$$= \sum_{\substack{p \in P'}} F_{t,p} + \sum_{\substack{s \in S}} F_{t,s} + \sum_{\substack{n \in N}} f_{t,n} \tag{6.14}$$

where $t \in T$ and $k \in K$. For every treatment unit, the inequality constraints are usually imposed upon the water throughput and the pollutant concentrations at the inlet, that is,

$$FT_t \leq FT_t^{\text{max}} \tag{6.15}$$

$$C_{t,k}^{\text{in}} \leq C_{t,k}^{\text{in,max}} \tag{6.16}$$

6.3.7 NEW TREATMENT UNITS

The model constraints for these new treatment units are primarily the same as those for the existing ones, that is,

$$C_{x,k}^{\text{in}}\left(1-R_{x,k}\right)=C_{x,k}^{\text{out}} \tag{6.17}$$

$$ft_x = \sum_{\substack{m\in M \\ m\neq x}} f_{m,x} = \sum_{\substack{n\in N \\ n\neq x}} f_{x,n} \tag{6.18}$$

$$ft_x C_{x,k}^{\text{in}} = \sum_{\substack{m\in M \\ m\neq x}} f_{m,x}C_{m,k} \tag{6.19}$$

where $x \in X$ and $k \in K$. For every new treatment unit, the upper bounds of the throughput and the pollutant concentrations at the inlet must also be included in the model, that is,

$$ft_x \leq ft_x^{\text{max}} \tag{6.20}$$

$$C_{x,k}^{\text{in}} \leq C_{x,k}^{\text{in,max}} \tag{6.21}$$

6.4   UNCERTAIN MULTIPLIERS

Because the actual operating conditions may vary with time, the values of some model parameters may be uncertain. A water network designed solely by nominal conditions may not be flexible enough to cope with all possible changes during operation. In this work, the following uncertain multipliers are adopted to facilitate systematic flexibility analysis:

$$FT_{w_1}^{\text{max}} = \overline{FT}_{w_1}^{\text{max}}\theta_{FT_{w_1}^{\text{max}}} \qquad \forall w_1 \in W_1 \tag{6.22}$$

$$FT_{w_2} = \overline{FT}_{w_2}\theta_{FT_{w_2}} \qquad \forall w_2 \in W_2 \tag{6.23}$$

$$C_{w_2,k} = \overline{C}_{w_2,k}\theta_{C_{w_2,k}} \qquad \forall w_2 \in W_2 \;\; \forall k \in K \tag{6.24}$$

$$ML_{u,k} = \overline{ML}_{u,k}\theta_{ML_{u,k}} \qquad \forall u \in U \;\; \forall k \in K \tag{6.25}$$

$$C_{u,k}^{\text{in,max}} = \overline{C}_{u,k}^{\text{in,max}}\theta_{C_{u,k}^{\text{in,max}}} \qquad \forall u \in U \;\; \forall k \in K \tag{6.26}$$

$$C_{u,k}^{\text{out,max}} = \overline{C}_{u,k}^{\text{out,max}}\theta_{C_{u,k}^{\text{out,max}}} \qquad \forall u \in U \;\; \forall k \in K \tag{6.27}$$

$$R_{t,k} = \overline{R}_{t,k}\theta_{R_{t,k}} \qquad \forall t \in T \;\; \forall k \in K \tag{6.28}$$

$$FT_t^{\text{max}} = \overline{FT}_t^{\text{max}}\theta_{FT_t^{\text{max}}} \qquad \forall t \in T \tag{6.29}$$

$$C_{t,k}^{\text{in,max}} = \overline{C}_{t,k}^{\text{in,max}}\theta_{C_{t,k}^{\text{in,max}}} \qquad \forall t \in T \;\; \forall k \in K \tag{6.30}$$

$$R_{x,k} = \bar{R}_{x,k}\theta_{R_{x,k}} \qquad \forall x \in X \quad \forall k \in K \tag{6.31}$$

$$ft_x^{max} = \bar{ft}_x^{max}\theta_{ft_x^{max}} \qquad \forall x \in X \tag{6.32}$$

$$C_{x,k}^{in,max} = \bar{C}_{x,k}^{in,max}\theta_{C_{x,k}^{in,max}} \qquad \forall x \in X \quad \forall k \in K \tag{6.33}$$

where $\overline{FT}_{w1}^{max}$, $\overline{FT}_{w2}$, $\bar{C}_{w2,k}$, $\overline{ML}_{u,k}$, $\bar{C}_{u,k}^{in,max}$, $\bar{C}_{u,k}^{out,max}$, $\bar{R}_{t,k}$, $\overline{FT}_t^{max}$, $\bar{C}_{t,k}^{in,max}$, $\bar{R}_{x,k}$, $\bar{ft}_x^{max}$, and $\bar{C}_{x,k}^{in,max}$ represent the nominal values of the uncertain parameters and $\theta_{FT_{w1}^{max}}$, $\theta_{FT_{w2}}$, $\theta_{C_{w2,k}}$, $\theta_{ML_{u,k}}$, $\theta_{C_{u,k}^{in,max}}$, $\theta_{C_{u,k}^{out,max}}$, $\theta_{R_{t,k}}$, $\theta_{FT_t^{max}}$, $\theta_{C_{t,k}^{in,max}}$, $\theta_{R_{x,k}}$, $\theta_{ft_x^{max}}$, and $\theta_{C_{x,k}^{in,max}}$ are the corresponding *uncertain multipliers*. Note that the nominal value of every uncertain multiplier always equals 1.

6.5 SINGLE-CONTAMINANT SYSTEMS

To facilitate clear explanation, let us first modify the previous model formulations for the single-contaminant systems by dropping the subscript k because it is only adopted to distinguish different contaminants.

6.5.1 Flexibility Assessment Strategies

Two alternative computation strategies can be taken to evaluate the steady-state flexibility indices of given water networks, and they are outlined next.

6.5.1.1 Flexibility Index Models Derived from the Active Set Method

Clearly, not all branches in the augmented superstructure are present in a given (revamp) design. The flow rates in the nonexistent branches should be set to zero by fixing all binary variables $d_{m,n}$ defined in Equation 6.1. The resulting inequality constraints established according to Equation 6.2 must then be used together with the material balances—that is, Equations 6.3 through 6.21, and the expected ranges of uncertain parameters, that is, Equations 6.22 through 6.33—to compute the flexibility index. In other words, these equality and inequality constraints should be substituted into Equations 2.15 through 2.25 to produce an MINLP model for flexibility assessment, and this model is referred to as the MINLP-FI model in the present chapter.

By utilizing the hyperbolic approximation technique proposed by Balakrishna and Biegler (1992), the smoothing function method (Raspanti et al., 2000) can be applied to convert the aforementioned MINLP model into an alternative nonconvex nonlinear program. Specifically, Equations 2.15 through 2.25 can be approximated with the following NLP model:

$$FI_s = \min_{\delta, \mu_i, \lambda_j, x_l, z_c, \theta_n} \delta \tag{6.34}$$

subject to the constraints in Equations 2.3, 2.4, and 2.17 through 2.19, 2.23, and 2.25 and

$$\lambda_j - \frac{1}{2}\left[\lambda_j + g_j + \sqrt{\left(\lambda_j + g_j\right)^2 + \varepsilon^2}\right] = 0 \quad \lambda_j \geq 0 \qquad (6.35)$$

Because the smoothing method is an approximation technique, this approach inevitably produces local solutions that may not be identical to those obtained with the original MINLP formulation. The validity of this approximation strategy depends to a great extent on the magnitude of the chosen parameter ε. A smaller ε yields a more accurate solution, but may cause ill conditioning. Because in general, the solution process of an NLP model converges faster, the corresponding results can be used at least as a good initial guess for solving the MINLP-FI model.

6.5.1.2 Flexibility Index Models Derived from the Vertex Method

As mentioned in the previous section, evaluating the flexibility index of water networks with the MINLP-FI model often requires elaborate initialization schemes and/or significant computation resources. It is therefore desirable to develop a more efficient solution approach based on the vertex method. Let us consider the corresponding mathematical formulation, that is, Equations 2.3, 2.4, and 2.26 through 2.28 and its solution procedure presented in Subsection 2.2.2. Notice that all vertices still have to be checked if this original version of the vertex method is to be implemented directly. However, it has also been observed by Chang et al. (2009) that the most constrained point (or the critical point) of a water network design can usually be associated with an upper or lower limit of each uncertain parameter by physical insights. These particular locations are (1) the upper bounds of (i) the mass loads of water-using units and (ii) the pollutant concentrations at the primary and secondary sources and (2) the lower bounds of (i) the removal ratios of wastewater treatment units, (ii) the allowed maximum inlet and outlet pollutant concentrations of water-using units, and (iii) the allowed maximum inlet pollutant concentration of wastewater treatment units. The flexibility index of a water network can thus be determined by this most constrained point alone. Such an improved model for computing flexibility index is referred to as the NLP-FI model in this chapter. It should be emphasized that the validity of this solution strategy has been confirmed empirically by extensive case studies in Chang et al. (2009).

6.5.2 RULE-BASED REVAMP STRATEGIES

6.5.2.1 Heuristic Rules

By the heuristic rules presented in the sequel, two classes of structural modifications can be considered to improve the operational flexibility of a given water network: (1) inserting/deleting pipeline connections and (2) adding/replacing treatment units. Although it is possible to automatically search for the needed design modification with mathematical programming models, such a brute force approach may not always be justifiable because the required computation load can grow exponentially, and we usually do not look for or need a network design in a revamping study that involves a drastic structural change. It is thus our intention to heuristically identify an acceptable solution while keeping the size of the optimization problem reasonably small.

For this purpose, several effective heuristics have been developed in this study by a large volume of test-case results (Riyanto, 2009). These heuristic design rules are briefly summarized here and, in general, they should be implemented in the same order given.

1. Introduce additional pipeline connection(s) to the nominal network so as to relax the active constraints in the solution of the original flexibility index model.

   a. Send clean water to the water-using unit to relax its inlet concentration constraint. This water can be taken from a water treatment unit or water-using unit.

   b. Increase the throughput of the water-using unit to relax its outlet concentration constraint. A higher throughput may be achieved with clean enough water from another unit.

   c. Send clean water to a water treatment unit to relax its inlet concentration constraint. Because the output of the treatment unit should be cleaner than its inputs, self-recycle is a viable option.

   d. Divert a portion of the water flow going into a sink to a water treatment unit(s) so as to relax the concentration constraint at the sink. This action must be considered along with other restrictions in the water network, such as the concentration and/or throughput constraints of the treatment unit(s) at the receiving end and the constraints of the unit(s) farther downstream.

2. Improve the performance of one or more existing treatment units. This task can be accomplished by replacing old units with better ones, by implementing a more up-to-date technology, by adding a post-treatment unit, or simply by repairing the existing treatment unit that is not working so well. It should also be noted that this approach does not guarantee flexibility enhancement.

3. Place one or more new treatment units to relax the active constraints described as follows:

   a. Place them before a water-using unit to relax its inlet and/or outlet concentration constraints.

   b. Put them in parallel with an existing unit to relieve its treatment load and to relax its active throughput constraint as well.

   c. Put them on the effluent flows to relax the concentration constraint at the sink.

Notice that these revamp measures can be roughly classified into three types based on capital investment costs, that is, (1) adding auxiliary pipelines, (2) upgrading existing treatment units, and (3) adding extra treatment units. Because their cost ranges are significantly different, these three options should be attempted sequentially one at a time according to the aforementioned order. If structural changes of the same type are being considered, they are ranked by their operating costs.

Three illustrative examples are presented to illustrate the implementation steps of these heuristic rules and to demonstrate the feasibility and benefits of the rule-based revamp procedure.

Example 6.2

Let us consider the nominal water network presented in Figure 6.3 (Riyanto and Chang, 2010). The corresponding model parameters and cost coefficients are shown in Table 6.2. The solution of the flexibility index model for this network is presented in Table 6.3. Let us assume that the upper limit of the freshwater supply rate cannot exceed 30 tonne/hr. Under this condition, Table 6.3 shows that the nominal network may not be resilient enough because the corresponding flexibility index is only 0.32. Due to budget constraints, it is also assumed in this example that new treatment units cannot be added to improve the system performance. From the values of the binary variables y_j, it is clear that the active constraints are associated with $\overline{F}_{w1}^U$, $\overline{F}_{t1}^U$, $\overline{F}_{t2}^U$, $\overline{Cl}_{u1}^U$, $\overline{CO}_{u1}^U$, $\overline{CO}_{u2}^U$, and $\overline{C}_{s1}^U$. Also it should be noted that only the last four of these active constraints may be relaxed with auxiliary pipelines. Let us first analyze those possibilities:

1. $\overline{Cl}_{u1}^U$: The inlet concentration of unit $u1$ cannot be lowered by adding new connections from other processing units, as the only flow with an acceptable concentration level, which should be lower than $\overline{Cl}_{u1}^U$ (1 ppm), is the freshwater (0.1 ppm).

2. $\overline{CO}_{u1}^U$: The corresponding constraint may be relaxed if $u1$ is operated at a higher throughput level. However, such a requirement cannot be satisfied with auxiliary pipelines because the constraint associated with $\overline{Cl}_{u1}^U$ is already active, and based on the argument against the revamp action mentioned earlier, it is not possible to find any secondary water source with a concentration lower than 1 ppm. Therefore, the present revamp option should be abandoned.

3. $\overline{CO}_{u2}^U$: It is feasible to relax this constraint because the inlet concentration of $u2$ in the nominal design does not reach its upper bound. A new connection is thus added from $u1$ to $u2$ as the water flow from $u1$ is the cleanest among all three water-using units and all treatment units have already been connected to $u2$. For convenience, this revamp option is referred to as

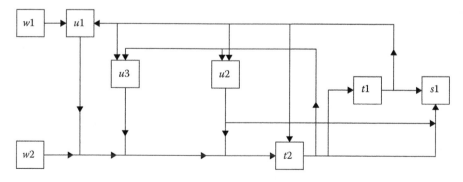

FIGURE 6.3 The nominal structure of the water network in Example 6.2. (Reprinted from *Chemical Engineering Science*, Vol. 65, Riyanto, E., Chang, C.T., A heuristic revamp strategy to improve operational flexibility of water networks based on active constraints, 2758–2770. Copyright 2010 with permission from Elsevier.)

design 6.2.A. It was determined that the flexibility index of design 6.2.A is the same as that of the original structure. Such a result is because the outlet stream of unit $u1$ is much dirtier than the inlet of unit $u2$.

4. $\overline{C}_{s1}^{U}$: None of the flows heading toward sink $s1$ can be diverted to the treatment units as they are all at their throughput limits.

As mentioned before, the second stage of the revamping procedure is to improve the separation efficiencies of existing treatment units. Let us consider two scenarios where the removal ratios of $t1$ and $t2$ can be enhanced to 0.95 (design 6.2.B) and 0.85 (design 6.2.C), respectively. Although the structures of both designs are the same as that of the nominal network, the corresponding flexibility indices can be raised, respectively, to 0.964 for design 6.2.B and 0.650 for design 6.2.C. This is because in these networks cleaner water can be produced with better treatment units and, consequently, cleaner inputs can be used in all

TABLE 6.2
Model Parameters Used in Example 6.2

Parameters		Values	Parameters	Value
$\overline{F}_{w1}^{U}$	(tonne/hour)	30.000	$\Delta\theta_{C_{w2}}^{+}$	0.100
$\overline{F}_{w2}$	(tonne/hour)	30	$\Delta\theta_{C_{w2}}^{-}$	0.100
$\overline{C}_{w1}$	(ppm)	0.100	$\Delta\theta_{M_{u1}}^{+}$	0.150
$\overline{C}_{w2}$	(ppm)	150.000	$\Delta\theta_{M_{u1}}^{-}$	0.150
$\overline{CI}_{u1}^{U}$	(ppm)	1.000	$\Delta\theta_{M_{u2}}^{+}$	0.150
$\overline{CI}_{u2}^{U}$	(ppm)	80.000	$\Delta\theta_{M_{u2}}^{-}$	0.150
$\overline{CI}_{u3}^{U}$	(ppm)	50.000	$\Delta\theta_{M_{u3}}^{+}$	0.150
$\overline{CI}_{t1}^{U}$	(ppm)	185.000	$\Delta\theta_{M_{u3}}^{-}$	0.150
$\overline{CI}_{t2}^{U}$	(ppm)	200.000	$\Delta\theta_{RR_{t1}}^{+}$	0.030
$\overline{CI}_{u1}^{U}$	(ppm)	101.000	$\Delta\theta_{RR_{t1}}^{-}$	0.030
$\overline{CI}_{u2}^{U}$	(ppm)	240.000	$\Delta\theta_{RR_{t2}}^{+}$	0.030
$\overline{CI}_{u3}^{U}$	(ppm)	200.000	$\Delta\theta_{RR_{t2}}^{-}$	0.030
$\overline{C}_{s1}$	(ppm)	10.000		
$\overline{F}_{t1}^{U}$	(tonne/hour)	125.000		
$\overline{F}_{t2}^{U}$	(tonne/hour)	135.000		
$\overline{M}_{u1}$	(kg/hour)	4.000		Cost coefficients
$\overline{M}_{u2}$	(kg/hour)	5.600	γ_{w1}	($/tonne)	1
$\overline{M}_{u3}$	(kg/hour)	4.500	γ_{t1}	($/tonne)	2
$\overline{RR}_{t1}$		0.9	γ_{t2}	($/tonne)	1
$\overline{RR}_{t2}$		0.8	γ_{s1}	($/tonne)	0

Source: Reprinted from *Chemical Engineering Science*, Vol. 65, Riyanto, E., Chang, C.T., A heuristic revamp strategy to improve operational flexibility of water networks based on active constraints, 2758–2770. Copyright 2010 with permission from Elsevier.

TABLE 6.3

A Solution of the MINLP-FI Model for Nominal Design in Example 6.2

Results		Values	Results	Values
$f_{w1,u1}$	(tonne/hr)	30.000	$y^U_{F_{w1}}$	1
$f_{w2,s2}$	(tonne/hr)	30.000	$y^U_{CI_{u1}}$	1
$f_{u1,t2}$	(tonne/hr)	41.921	$y^U_{CI_{u2}}$	0
$f_{u2,t2}$	(tonne/hr)	24.116	$y^U_{CI_{u3}}$	0
$f_{u2,s1}$	(tonne/hr)	0.676	$y^U_{CI_{t1}}$	0
$f_{u3,t2}$	(tonne/hr)	38.963	$y^U_{CI_{t2}}$	0
$f_{t1,u1}$	(tonne/hr)	11.921	$y^U_{CO_{u1}}$	1
$f_{t1,u2}$	(tonne/hr)	24.791	$y^U_{CO_{u2}}$	1
$f_{t1,u3}$	(tonne/hr)	38.079	$y^U_{CO_{u3}}$	0
$f_{t1,s1}$	(tonne/hr)	50.208	$y^U_{C_{s1}}$	1
$f_{t2,u3}$	(tonne/hr)	0.883	$y^U_{F_{t1}}$	1
$f_{t2,t1}$	(tonne/hr)	125.000	$y^U_{F_{t2}}$	1
$f_{t2,s1}$	(tonne/hr)	9.117		
CI_{u1}	(ppm)	1.000		
CI_{u2}	(ppm)	3.265		
CI_{u3}	(ppm)	3.872		
CI_{t1}	(ppm)	30.050		
CI_{t2}	(ppm)	144.689		
CO_{u1}	(ppm)	101.000		
CO_{u2}	(ppm)	240.000		
CO_{u3}	(ppm)	124.916		
CO_{t1}	(ppm)	3.265		
CO_{t2}	(ppm)	30.050	Minimized operation cost ($/hr)	266.137
C_{s1}	(ppm)	10.000	Flexibility index	0.320

Source: Reprinted from *Chemical Engineering Science*, Vol. 65, Riyanto, E., Chang, C.T., A heuristic revamp strategy to improve operational flexibility of water networks based on active constraints, 2758–2770. Copyright 2010 with permission from Elsevier.

water-using units and sinks. As a result, the constraints associated with $\overline{CI}^U_{u1}$, $\overline{CO}^U_{u1}$, $\overline{CO}^U_{u2}$, and $\overline{C}^U_{s1}$ can be relaxed simultaneously.

Next let us consider the possibility of augmenting design 6.2.B or design 6.2.C with additional auxiliary pipelines. According to the earlier discussions, we can see that it is only possible to relax the active constraint associated with $\overline{CO}^U_{u2}$ in the original model by adding auxiliary pipelines. Because the network structures of design 6.2.B and 6.2.C are identical to that of the nominal design and there are no new active constraints, it can be expected that the only candidate constraint is again associated with $\overline{CO}^U_{u2}$. Because this constraint is not active in design 6.2.B, it is only necessary to evaluate the benefit of adding a connection from $u1$ to $u2$ in design 6.2.C (which will be referred to as design 6.2.D). It can be observed that the flexibility levels of design 6.2.C and 6.2.D are the same. The argument applied to explain why design 6.2.A fails to achieve a better performance over the nominal design—that is, the positive effect of increasing throughput of unit $u2$ is canceled out by the increase in inlet concentration—is also applicable in the present case.

Our last resort (design 6.2.E) is a combination of design 6.2.B and design 6.2.C. The optimization results show that the corresponding flexibility index can be improved to 1.478. This indicates that design 6.2.E should be a suitable revamp candidate for the present application.

Example 6.3

Let us consider the water network presented in Figure 6.4. The model parameters and the cost coefficients adopted in the present case are the same as those used previously in Example 6.2 (see Table 6.2). Let us assume that the upper limits of the freshwater supply rate also cannot exceed 30 tonne/hr. The optimal solution of the corresponding flexibility index model can be found in Table 6.4. Notice that this nominal network is not flexible enough because FI_s reaches only 0.387. The active constraints in this case are those associated with $\bar{F}_{w1}^U$, $\bar{F}_{t1}^U$, $\bar{F}_{t2}^U$, $\overline{CO}_{u1}^U$, $\overline{CO}_{u2}^U$, $\overline{CO}_{u3}^U$, and $\bar{C}_{s1}^U$, whereas only the last four may be relaxed by adding auxiliary pipelines. Let us evaluate these possibilities first.

1. $\overline{CO}_{u1}^U$, $\overline{CO}_{u2}^U$, and $\overline{CO}_{u3}^U$: The active constraint associated with the outlet concentration of a water-using unit can often be relaxed by introducing an additional clean water flow to increase its throughput. The water flows from treatment units cannot be selected for this purpose, as all of them are used to maintain the active constraints on other units and/or sinks. Specifically, the output from $t1$ is needed to keep the constraints for $\overline{CO}_{u1}^U$, $\overline{CO}_{u2}^U$, $\overline{CO}_{u3}^U$, and $\bar{C}_{s1}^U$, whereas the output from $t2$ is necessary for $\overline{CO}_{u1}^U$, $\overline{CO}_{u2}^U$, and $\bar{C}_{s1}^U$. Sharing these flows will inevitably result in a lower flexibility level. Thus, the output from unit $u1$ is chosen to be the water source for increasing the throughputs of $u2$ and $u3$ because its concentration is the lowest among

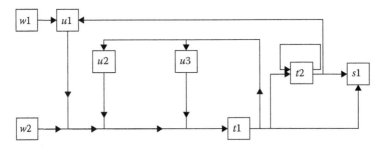

FIGURE 6.4 The nominal structure of the water network in Example 6.3. (Reprinted from *Chemical Engineering Science*, Vol. 65, Riyanto, E., Chang, C.T., A heuristic revamp strategy to improve operational flexibility of water networks based on active constraints, 2758–2770. Copyright 2010 with permission from Elsevier.)

TABLE 6.4

A Solution of the MINLP-FI Model for the Nominal Design in Example 6.3

Results		Values	Results	Values
$f_{w1,u1}$	(tonne/hr)	30.000	$y_{F_{w1}}^U$	1
$f_{w2,t1}$	(tonne/hr)	30.000	$y_{CI_{u1}}^U$	0
$f_{u1,t1}$	(tonne/hr)	42.098	$y_{CI_{u2}}^U$	0
$f_{u2,t1}$	(tonne/hr)	26.713	$y_{CI_{u3}}^U$	0
$f_{u3,t1}$	(tonne/hr)	26.188	$y_{CI_{t1}}^U$	0
$f_{t1,u2}$	(tonne/hr)	26.713	$y_{CI_{t2}}^U$	0
$f_{t1,u3}$	(tonne/hr)	26.188	$y_{CO_{u1}}^U$	1
$f_{t1,t2}$	(tonne/hr)	41.284	$y_{CO_{u2}}^U$	1
$f_{t1,s1}$	(tonne/hr)	30.814	$y_{CO_{u3}}^U$	1
$f_{t2,u1}$	(tonne/hr)	12.098	$y_{C_{s1}}^U$	1
$f_{t2,t2}$	(tonne/hr)	93.716	$y_{F_{t1}}^U$	1
$f_{t2,s1}$	(tonne/hr)	29.186	$y_{F_{t2}}^U$	1
CI_{u1}	(ppm)	0.463		
CI_{u2}	(ppm)	18.182		
CI_{u3}	(ppm)	18.182		
CI_{t1}	(ppm)	164.601		
CI_{t2}	(ppm)	6.505		
CO_{u1}	(ppm)	101.000		
CO_{u2}	(ppm)	240.000		
CO_{u3}	(ppm)	200.000		
CO_{t1}	(ppm)	18.182		
CO_{t2}	(ppm)	1.362	Minimized operation cost ($/hr)	307.265
C_{s1}	(ppm)	10.000	Flexibility index	0.387

Source: Reprinted from *Chemical Engineering Science*, Vol. 65, Riyanto, E., Chang, C.T., A heuristic revamp strategy to improve operational flexibility of water networks based on active constraints, 2758–2770. Copyright 2010 with permission from Elsevier.

all eligible candidates. Also, unit $u1$ is a bad choice to be considered as the recipient of additional flow because its inlet concentration limit is very low (1 ppm). A revamp design is generated by adding a pipeline from unit $u1$ to unit $u2$ (design 6.3.A), and another design by adding the connection from $u1$ to $u3$ (design 6.3.B). The optimal solutions of the corresponding flexibility index models show that these design options raise FI_s to 1.150 for design 6.3.A and to 0.658 for design 6.3.B. In the former case, although the revamp action significantly causes an increase in the inlet concentration of unit $u2$, the active constraint corresponding to $\overline{CO}_{u2}^U$ is relaxed due to the increase of throughput in unit $u2$ (42.340 tonne/hour in design 6.3.A versus 26.713 tonne/hour in the nominal design). On the other hand, the flexibility level of design 6.3.B is improved significantly, although not enough to reach the desired level of 1. It appears that although the constraint for $\overline{CO}_{u3}^{U}$ is relaxed, the constraint associated with $\overline{CO}_{u2}^{U}$ is still active. To confirm that the exit flow from unit $u1$ is indeed the best candidate for use as the needed additional water source, we have tried to determine the effects of connecting unit $u3$ to $u2$ (design 6.3.C). The solution of the corresponding model shows that the flexibility index of design 6.3.C (0.535) is smaller than those achieved in designs 6.3.A and 6.3.B. The optimal solution from the corresponding flexibility index model also shows that the active constraints in all three cases are the same, except an extra one—that associated with $\overline{Cl}_{u2}^{U}$ —is embedded in design 6.3.C. This finding reveals that although adding a connection from $u3$ to $u2$ is capable of relaxing the constraint for $\overline{CO}_{u2}^{U}$, the constraint for $\overline{Cl}_{u2}^{U}$ becomes a new bottleneck, which prevents design 6.3.C from further improving its flexibility level. This is not a problem in design 6.3.A because the pollutant concentration of the water flow from unit $u1$ is much lower than that from $u3$.

2. $\overline{C}_{s1}^{U}$: The water flows heading toward sink $s1$ cannot be diverted to the treatment units as they are at their throughput limits. Because in this case it has already been shown that design 6.3.A is capable of compensating for the anticipated disturbances, it is not necessary to further consider upgrading/replacing treatment units. Therefore, our final selection in this example should be design 6.3.A.

Example 6.4

Let us consider the nominal water network presented in Figure 6.5, which consists of one freshwater source, one secondary water source, four water-using units, one treatment unit, and one sink. In addition to the uncertain

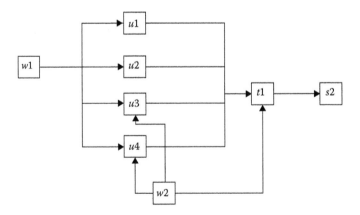

FIGURE 6.5 The nominal structure of the water network in Example 6.4. (Reprinted from *Chemical Engineering Science*, Vol. 65, Riyanto, E., Chang, C.T., A heuristic revamp strategy to improve operational flexibility of water networks based on active constraints, 2758–2770. Copyright 2010 with permission from Elsevier.)

parameters discussed in the previous two examples, the random disturbances in the freshwater supply rate ($\theta_{F_{w1}^U}$) and freshwater quality ($\theta_{C_{w1}^U}$) are considered here. It is assumed that the upper limit of the freshwater supply rate cannot exceed 25 tonne/hr and all revamp options are allowed in this example. The corresponding model parameters and cost coefficients are shown in Table 6.5, and the solution of the flexibility index model for this network can be found in Table 6.6. It can be observed that the flexibility index of the nominal design is only 0.249, and from the values of the binary variables y_j, the active constraints are associated with $\overline{F}_{w1}^U$, $\overline{CI}_{u4}^U$, $\overline{CO}_{u1}^U$, $\overline{CO}_{u2}^U$, $\overline{CO}_{u3}^U$, and $\overline{CO}_{u4}^U$. According to the proposed heuristics, the last five of these active constraints may be relaxed with auxiliary pipelines. Let us examine these possibilities.

1. $\overline{CI}_{u4}^U$ and $\overline{CO}_{u4}^U$: Both constraints can be relaxed by increasing the throughput and/or lowering the inlet concentration of unit $u4$. These

TABLE 6.5

Model Parameters Used in Example 6.4

Parameters		Values	Parameters	Value	
$\overline{F}_{w1}^U$	(tonne/hour)	25	$\Delta\theta_{F_{w1}^U}^+$	0.100	
$\overline{F}_{w2}$	(tonne/hour)	100	$\Delta\theta_{F_{w1}^U}^-$	0.100	
$\overline{C}_{w1}$	(ppm)	0.050	$\Delta\theta_{C_{w1}}^-$	0.100	
$\overline{C}_{w2}$	(ppm)	100.000	$\Delta\theta_{C_{w1}}^-$	0.100	
$\overline{CI}_{u1}^U$	(ppm)	1.000	$\Delta\theta_{C_{w2}}^+$	0.100	
$\overline{CI}_{u2}^U$	(ppm)	50.000	$\Delta\theta_{C_{w2}}^-$	0.050	
$\overline{CI}_{u3}^U$	(ppm)	100.000	$\Delta\theta_{M_{u1}}^+$	0.150	
$\overline{CI}_{u4}^U$	(ppm)	100.000	$\Delta\theta_{M_{u1}}^-$	0.150	
$\overline{CI}_{t1}^U$	(ppm)	200.000	$\Delta\theta_{M_{u2}}^+$	0.150	
$\overline{CO}_{u1}^U$	(ppm)	50.000	$\Delta\theta_{M_{u2}}^-$	0.150	
$\overline{CO}_{u2}^U$	(ppm)	250.000	$\Delta\theta_{M_{u3}}^+$	0.150	
$\overline{CO}_{u3}^U$	(ppm)	200.000	$\Delta\theta_{M_{u3}}^-$	0.150	
$\overline{CO}_{u4}^U$	(ppm)	200.000	$\Delta\theta_{M_{u4}}^+$	0.150	
$\overline{C}_{s1}$	(ppm)	50.000	$\Delta\theta_{M_{u4}}^-$	0.150	
$\overline{F}_{t1}^U$	(tonne/hour)	125.000			
$\overline{M}_{u1}$	(kg/hour)	0.100			
$\overline{M}_{u2}$	(kg/hour)	2.000	Cost coefficients		
$\overline{M}_{u3}$	(kg/hour)	5.000	γ_{w1}	($/tonne)	2
$\overline{M}_{u4}$	(kg/hour)	7.000	γ_{t1}	($/tonne)	1
$\overline{RR}_{t1}$		0.8	γ_{s1}	($/tonne)	1.5

Source: Reprinted from *Chemical Engineering Science*, Vol. 65, Riyanto, E., Chang, C.T., A heuristic revamp strategy to improve operational flexibility of water networks based on active constraints, 2758–2770. Copyright 2010 with permission from Elsevier.

TABLE 6.6

A Solution of the MINLP-FI Model for the Nominal Design in Example 6.4

Results		Values	Results	Values
$f_{w1,u1}$	(tonne/hr)	2.077	$y_{F_{w1}}^U$	1
$f_{w1,u2}$	(tonne/hr)	8.300	$y_{CI_{u1}}^U$	0
$f_{w1,u3}$	(tonne/hr)	12.239	$y_{CI_{u2}}^U$	0
$f_{w1,u4}$	(tonne/hr)	1.762	$y_{CI_{u3}}^U$	0
$f_{w2,u3}$	(tonne/hr)	28.090	$y_{CI_{u4}}^U$	1
$f_{w2,u4}$	(tonne/hr)	70.848	$y_{CI_{t1}}^U$	1
$f_{w2,t1}$	(tonne/hr)	1.061	$y_{CO_{u1}}^U$	1
$f_{u1,t1}$	(tonne/hr)	2.077	$y_{CO_{u2}}^U$	1
$f_{u2,t1}$	(tonne/hr)	8.300	$y_{CO_{u3}}^U$	1
$f_{u3,t1}$	(tonne/hr)	40.330	$y_{CO_{u4}}^U$	1
$f_{u4,t1}$	(tonne/hr)	72.610	$y_{C_{s1}}^U$	0
$f_{t1,s1}$	(tonne/hr)	124.378	$y_{F_{t1}}^U$	0
CI_{u1}	(ppm)	0.051		
CI_{u2}	(ppm)	0.051		
CI_{u3}	(ppm)	71.399		
CI_{u4}	(ppm)	100.000		
CI_{t1}	(ppm)	200.000		
CO_{u1}	(ppm)	50.000		
CO_{u2}	(ppm)	250.000		
CO_{u3}	(ppm)	200.000		
CO_{u4}	(ppm)	200.000		
CO_{t1}	(ppm)	40.000	Minimized operation cost ($/hr)	342.273
C_{s1}	(ppm)	40.000	Flexibility index	0.249

Source: Reprinted from *Chemical Engineering Science*, Vol. 65, Riyanto, E., Chang, C.T., A heuristic revamp strategy to improve operational flexibility of water networks based on active constraints, 2758–2770. Copyright 2010 with permission from Elsevier.

tasks can be achieved by sending a clean water flow to $u4$. According to Table 6.6, the output of unit $t1$ is clearly the best choice because its pollutant concentration is relatively low and adding a stream from $t1$ to $u4$ will not affect the other constraints. Therefore, we connect unit $t1$ to unit $u4$ as our first design option (design 6.4.A). The optimal solution of the corresponding MINLP-FI model shows that the flexibility index of design 6.4.A can only be raised to 0.280. This is because the output of $u4$ is eventually directed to $t1$ and the throughput limit of unit $t1$ prevents a large recycle flow from $t1$ to $u4$.

2. $\overline{Cl}_{t1}^{U}$: This constraint may be relaxed by adding a self-recycling flow around unit $t1$ as the next design option (design 6.4.B). From the optimal solution of the corresponding MINLP-FI model, it can be observed that the flexibility index again can only reach 0.280. The reason for this is the same as that described previously for design 6.4.A.

3. $\overline{CO}_{u1}^{U}$: Because the inlet concentration limit of $u1$ is very low, it is not possible to identify a water source in the given network that is clean enough for relaxing the constraint under consideration.

4. $\overline{CO}_{u2}^{U}$ and $\overline{CO}_{u3}^{U}$: These two constraints may be relaxed by diverting the outlet stream from $t1$ to $u2$ and $u3$, respectively, to increase their throughputs. This action obviously results in the same problem as that encountered in design 6.4.A and 6.4.B—namely, the throughput limit of unit $t1$. To confirm this prediction, the impacts of connecting unit $t1$ to unit $u2$ (design 6.4.C) and connecting unit $t1$ to unit $u3$ (design 6.4.D) have been evaluated. As expected, neither design 6.4.C nor 6.4.D can be adopted to improve the operational flexibility to a satisfactory level. In fact, the flexibility levels in both cases are the same as those achieved in design 6.4.A and design 6.4.B. Because each of the two options is individually hampered by the throughput limit of unit $u1$, the improvement should still be minimal if both changes are combined in the next design option, design 6.4.E. This prediction can be confirmed in the solution of the corresponding MINLP-FI model, as the flexibility index for this design is also 0.280.

In the next phase of the proposed revamp procedure, it is required to upgrade the existing treatment unit (design 6.4.F). Let us assume that the removal ratio of unit $t1$ can be improved to 0.9. However, because the water flow from unit $t1$ is directed only to the sink in the nominal design, upgrading $t1$ only relaxes the concentration constraint at $s1$. As there are no changes in the network structure, no improvement can be anticipated either. The solution of the corresponding model shows that the flexibility level of design 6.4.F is not different from the original level (0.249). Therefore, it can be concluded that the operational flexibility of a water network cannot be enhanced by applying the design heuristics to improve a constraint that is originally not active (corresponding to $\overline{C}_{s1}^{U}$ in this particular case).

In the final revamp phase, the possibilities of installing additional treatment units are explored. Let us assume that the available new treatment units are of the same type and their removal ratios are the same (0.9). Following is a list of possible locations for these units:

1. $\overline{Cl}_{u4}^{U}$ and $\overline{CO}_{u4}^{U}$: These two constraints can be relaxed simultaneously by placing the new treatment unit $t2$ before unit $u4$ to lower the pollutant concentration of the secondary water and by diverting a portion of the outlet flow of unit $t2$ to sink $s1$. This option is referred to as design 6.4.G (see Figure 6.6). The solution of the corresponding model shows that this design is flexible enough, as the flexibility index can be improved to 1.604.

2. $\overline{CO}_{u1}^{U}$ and $\overline{CO}_{u2}^{U}$: As freshwater is used in $u1$ and $u2$, it makes no sense to install new treatment units to produce cleaner inputs for these units.

3. $\overline{CO}_{u3}^{U}$: This constraint can be relaxed by installing a new treatment unit $t2$ before $u3$ to lower the pollutant concentration of the secondary water source and again diverting a portion of the outlet flow of $t2$ to sink $s1$ (see design 6.4.H in Figure 6.7). The rationale for adopting this design is similar to that for design 6.4.G. The solution of the corresponding flexibility index model shows that the operational flexibility of design 6.4.H also reaches a satisfactory level of 1.314.

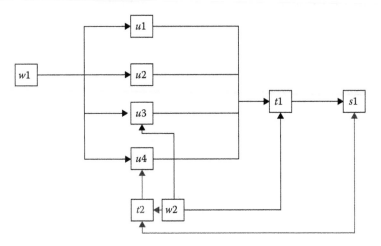

FIGURE 6.6 Revamp design 6.4.G. (Reprinted from *Chemical Engineering Science*, Vol. 65, Riyanto, E., Chang, C.T., A heuristic revamp strategy to improve operational flexibility of water networks based on active constraints, 2758–2770. Copyright 2010 with permission from Elsevier.)

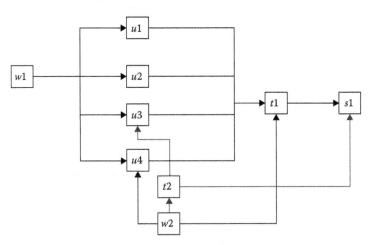

FIGURE 6.7 Revamp design 6.4.H. (Reprinted from *Chemical Engineering Science*, Vol. 65, Riyanto, E., Chang, C.T., A heuristic revamp strategy to improve operational flexibility of water networks based on active constraints, 2758–2770. Copyright 2010 with permission from Elsevier.)

4. $\bar{F}_{t1}^{U}$: Although this constraint is not active in the original nominal design, it becomes active in several revamped versions, that is, 6.4.A, 6.4.B, 6.4.C, 6.4.D, and 6.4.E. Let us try to improve design 6.4.B by installing unit $t2$ to work in parallel with unit $t1$, thus relaxing its throughput limit (see design 6.4.I in Figure 6.8). The optimal solution shows that this design can also be adapted to adequately compensate for the anticipated disturbances, as the flexibility index is raised to 1.071.

Because more than one design—that is, 6.4.G, 6.4.H, and 6.4.I—can be used to achieve the desired flexibility level, additional criteria (e.g., total capital investment and operating cost) must be adopted to select the most appropriate one for the actual application. A more detailed discussion can be found in Riyanto and Chang (2010).

6.5.3 MODEL-GUIDED REVAMP STRATEGIES

6.5.3.1 Extra Model Constraints

Other than the model constraints listed in Section 6.3, additional ones are needed to facilitate the formulation of the utility models for generating the revamp designs.

1. Overdesign levels:
   If the freshwater supply rate can be adequately controlled, its upper bound should not be treated as an uncertain parameter, that is, $\theta_{FT_{w1}^{max}} = 1$. Thus, FT_{w1}^{max} in Equation 6.4 can be viewed as the chosen capacity of the freshwater supply system. The overdesign level in this capacity can be expressed as

$$FT_{w1}^{max} = \overline{SC}_{w1}\left(1+O_{w1}^{S}\right) \tag{6.36}$$

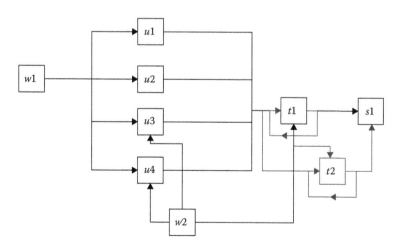

FIGURE 6.8 Revamp design 6.4.I, (Reprinted from *Chemical Engineering Science*, Vol. 65, Riyanto, E., Chang, C.T., A heuristic revamp strategy to improve operational flexibility of water networks based on active constraints, 2758–2770, Copyright 2010 with permission from Elsevier.)

where $\overline{SC}_{w1}$ and O_{w1}^S, respectively, represent the nominal supply rate and the corresponding overdesign percentage of freshwater w_1, and both should be considered as given parameters. On the other hand, the upper limit of the flow rate in each existing pipeline can be written as

$$F_{m,n} \leq \overline{FC}_{m,n}\left(1+O_{m,n}^F\right) \quad m \in M, n \in N \tag{6.37}$$

where $\overline{FC}_{m,n}$ is the nominal flow rate of the existing stream from split node m to mixing node n, and $O_{m,n}^F$ is the corresponding overdesign level.

2. Critical direction:

This critical direction in the parameter space is determined with the upper or lower limit of each uncertain parameter by physical insights. Specifically, this direction should be associated with the following:

a. The upper bounds of the pollutant concentrations at secondary sources and the mass loads of water-using units, that is,

$$\theta_{C_{w2}} = 1+\delta\Delta\theta_{C_{w2}}^+ \quad \forall w_2 \in W_2 \tag{6.38}$$

$$\theta_{ML_u} = 1+\delta\Delta\theta_{ML_u}^+ \quad \forall u \in U \tag{6.39}$$

b. The lower bounds of the allowed maximum inlet and outlet pollutant concentrations of water-using units, the removal ratios of wastewater treatment units, the allowed maximum throughputs of treatment units, and the allowed maximum inlet pollutant concentrations of treatment units, that is,

$$\theta_{C_u^{\text{in,max}}} = 1-\delta\Delta\theta_{C_u^{\text{in,max}}}^- \quad \forall u \in U \tag{6.40}$$

$$\theta_{C_u^{\text{out,max}}} = 1-\delta\Delta\theta_{C_u^{\text{out,max}}}^- \quad \forall u \in U \tag{6.41}$$

$$\theta_{R_t} = 1-\delta\Delta\theta_{R_t}^- \quad \forall t \in T \tag{6.42}$$

$$\theta_{R_x} = 1-\delta\Delta\theta_{R_x}^- \quad \forall x \in X \tag{6.43}$$

$$\theta_{FT_t^{\text{max}}} = 1-\delta\Delta\theta_{FT_t^{\text{max}}}^- \quad \forall t \in T \tag{6.44}$$

$$\theta_{ft_x^{\text{max}}} = 1-\delta\Delta\theta_{ft_x^{\text{max}}}^- \quad \forall x \in X \tag{6.45}$$

$$\theta_{C_t^{\text{in,max}}} = 1-\delta\Delta\theta_{C_t^{\text{in,max}}}^- \quad \forall t \in T \tag{6.46}$$

$$\theta_{C_x^{\text{in,max}}} = 1-\delta\Delta\theta_{C_x^{\text{in,max}}}^- \quad \forall x \in X \tag{6.47}$$

Finally, note that the critical limit for the supply rate of every secondary source can only be identified on a case-by-case basis. If the secondary water is too dirty to be consumed by any water-using unit, the upper bound of its flow rate should be treated as the limiting constraint. Otherwise, the lower bound must be chosen.

3. Nonexistent flows:

The nonexistent branches of the given network may or may not be selected as the auxiliary pipeline for flexibility enhancement. The selected ones should be constrained by fixing the binary parameters in Equation 6.2 to be 1, whereas the remaining parameters must be set to 0.

6.5.3.2 Utility Models

If a given nominal network is infeasible, the utility models presented next can be used to determine the exact (lowest) overdesign level of a freshwater supply system and/or to identify the optimal structural changes so as to cope with all possible variations defined in the expected region of uncertain parameters, that is, when $\delta = 1$ in Equation 2.10. Their formulations are summarized in the sequel.

- *Minimal Source Capacity.* The utility model for calculating the smallest upper limit of the total freshwater supply rate is referred to as the NLP-SC model. The model formulation can be expressed as

$$\min \sum_{w_1 \in W_1} FT_{w_1} \qquad (6.48)$$

subject to Equations 6.1 through 6.33, 6.35 through 6.46, and

$$\delta = 1 \qquad (6.49)$$

Notice that because usually there is only one primary source, the minimized objective value in this case can also be used to determine the desired overdesign level of the freshwater supply system.

- *Optimal Network Reconfiguration.* As mentioned before, the network configuration can be modified by adding new pipelines and/or removing existing ones. In principle, these pipelines are selected mainly to relax one or more active constraints so as to create chances for further flexibility increase (Riyanto and Chang, 2010). To facilitate construction of a mathematical programming model to reconfigure the network connections automatically, the following inequality constraints must be imposed upon the flow rates that are facilitated with replaced pipelines:

$$F_{m,n} \leq y_{m,n} F^U \qquad m \in M, n \in N \qquad (6.50)$$

where F^U is a large enough positive constant and $F^U > \overline{FC}_{m,n}(1 + O^F_{m,n})$; $y_{m,n}$ is a binary variable used to signify whether or not the corresponding pipeline can be replaced in the final network design. On the other hand, the flow rate in a selected new connection, that is, when $d_{m',n'} = 1$, must also be constrained according to

$$f_{m',n'} \leq z_{m',n'} f^U \qquad m' \in M, n' \in N \tag{6.51}$$

where f^U is a design parameter defined in Equation 6.2 and $z_{m',n'}$ is a binary variable used to signify whether or not the corresponding pipeline can be added in the final network design. To minimize the total capital expenditure, the following simplified objective function is used in the utility model for optimal network reconfiguration:

$$\min \left[C_{pl} \left(\sum_{\substack{m \in M \\ n \in N}} y_{m,n} + \sum_{\substack{m' \in M \\ n' \in N}} z_{m',n'} \right) + \sum_{w_1 \in W_1} C_{w_1} FT_{w_1} \right] \tag{6.52}$$

where C_{pl} is the average annualized cost of installing and operating a new pipeline and C_{w1} is the model coefficient for the annualized capital cost of the freshwater supply system $w1$. It should be noted that a more elaborate cost model can certainly be adopted if accurate cost data are available. Notice also that the model constraints in this mathematical program have already been described in Equations 6.1 through 6.33, 6.35 through 6.46, and 6.48, and it is referred to as the MINLP-NR model in this chapter.

The use of utility models for generating revamp designs is illustrated with an example given next.

Example 6.5

Let us consider the grassroots design problem studied in Chang et al. (2009). There are two water sources, three water-using units, two wastewater treatment units, and a wastewater sink in this chemical process. The nominal flow rate and contaminant concentration of the water sources are presented in Table 6.7, and the design specifications of the water-using units and wastewater treatment units are provided in Tables 6.8 and 6.9, respectively. Finally, the pollutant concentration at the sink is required to be kept below 10 ppm.

TABLE 6.7
Nominal Stream Data of Water Sources in Example 6.5

	$\bar{F}^W$ (ton/hr)	$\bar{C}_w$ (ppm)
$w1$	-	0.1
$w2$	30	150.0

Source: Reprinted with permission from Li and Chang 2011, 3763–3774. Copyright 2011 American Chemical Society.

By minimizing the freshwater consumption rate by a conventional superstructure, two alternative designs can be generated with the aforementioned nominal data. For convenience, they are referred to as design 6.5.I (Figure 6.9) and design 6.5.II (Figure 6.10), respectively. Because the flow ratios of the streams branched from a splitter may be adjusted to compensate for external disturbances during operation, all splitters are marked in the figures by small circles to facilitate intuitive assessment of operational flexibility. The nominal operating conditions of the water-using and wastewater treatment units in these designs are provided in Tables 6.10 and 6.11. In this example, these two structures are used as the base-case designs for the subsequent flexibility analysis. Notice that although the numbers of branches (12) and splitters (5) are the same in both networks, the freshwater usage of design 6.5.I is 26.489 tonne/h, whereas much less (8.384 tonne/h) is needed in design 6.5.II. The reduction of the freshwater requirement is achieved in the latter case by allowing the self-recycle stream around treatment unit $t2$. The overdesign levels of the freshwater supply system and all pipelines in both designs are set at 30% and 50%, respectively. Therefore, the upper bound of the freshwater consumption rate should be 34.436 tonne/h in design 6.5.I, whereas it is 10.90 tonne/h in design 6.5.II.

TABLE 6.8

Nominal Design Specifications of Water-Using Units in Example 6.5

Unit	$\bar{C}^{in}_{max}$ (ppm)	$\bar{C}^{out}_{max}$ (ppm)	$\bar{F}^{in}_{lim}$ (ton/hr)	$\bar{M}$ (kg/hr)
$u1$	1	101	40	4.0
$u2$	80	240	35	5.6
$u3$	50	200	30	4.5

Source: Reprinted with permission from Li and Chang 2011, 3763–3774. Copyright 2011 American Chemical Society.

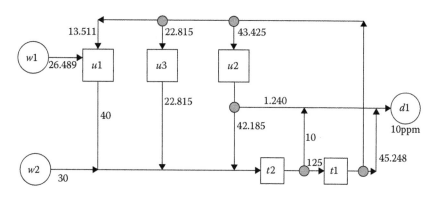

FIGURE 6.9 Design 6.5.I. (Reprinted with permission from Li and Chang 2011, 3763–3774. Copyright 2011 American Chemical Society.)

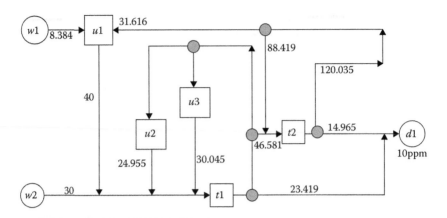

FIGURE 6.10   Design 6.5.II. (Reprinted with permission from Li and Chang 2011, 3763–3774. Copyright 2011 American Chemical Society.)

TABLE 6.9

Nominal Design Specifications of Wastewater Treatment Units in Example 6.5

Unit	$\bar{C}^{in}_{max}$ (ppm)	$\bar{F}^{in}_{max}$ (ton/hr)	Removal Ratio $\bar{R}$
$t1$	185	125	0.9
$t2$	200	135	0.8

Source: Reprinted with permission from Li and Chang 2011, 3763–3774. Copyright 2011 American Chemical Society.

TABLE 6.10

Nominal Operating Conditions of All Units in Design 6.5.I

Unit	$u1$	$u2$	$u3$	$t1$	$t2$	$d1$
Flow rate (ton/hr)	40.000	43.425	22.815	125.000	135.000	56.489
C^{in} (ppm)	1.000	2.764	2.764	27.644	138.220	10.000
C^{out} (ppm)	101.000	131.723	200.000	2.764	27.644	

Source: Reprinted with permission from Li and Chang 2011, 3763–3774. Copyright 2011 American Chemical Society.

Let us next assume in this example that the external disturbances during normal operation may cause three types of design parameters to fluctuate: (1) the contaminant concentration in secondary water, (2) the mass load of every water-using unit, and (3) the removal ratio of every wastewater treatment unit.

TABLE 6.11

Nominal Operating Conditions of All Units in Design 6.5.II

Unit	$u1$	$u2$	$u3$	$t1$	$t2$	$d1$
Flow rate (ton/hr)	40.000	24.955	30.045	125.000	135.000	38.384
C^{in} (ppm)	1.000	15.598	15.598	155.983	6.193	10.000
C^{out} (ppm)	101.000	240.000	165.375	15.598	1.239	

Source: Reprinted with permission from Li and Chang 2011, 3763–3774. Copyright 2011 American Chemical Society.

Thus, the following uncertain multipliers were introduced into the flexibility index models:

$$0.9 \leq \theta_{C_{w2}} \leq 1.1 \quad w_2 \in W_2 \tag{6.53}$$

$$0.85 \leq \theta_{ML_u} \leq 1.15 \quad u \in U \tag{6.54}$$

$$0.97 \leq \theta_{R_t}, \theta_{R_x} \leq 1.03 \quad t \in T \quad x \in X \tag{6.55}$$

Notice that these multipliers have already been defined in Section 6.4. A brief summary of the steps in generating the revamp designs is given here:

1. The flexibility indices of design 6.5.I and design 6.5.II can be found with the NLP-FI model to be 0.765 and 0.113, respectively. In both cases, the freshwater consumption rates at critical conditions reached their respective upper bounds. Thus, it is clear that the expected uncertain disturbances cannot be compensated for by adjusting the control variables in both cases. The subsequent assessment steps should then be applied to these two nominal designs individually.

2. Let us first consider design 6.5.I. The possibility of raising its operational flexibility by relaxing the upper bound of the freshwater supply rate is first explored. It was found by solving the NLP-FI model again that the flexibility index can be improved to 1.351 if this upper limit is increased to 40 tonne/h. Under the critical condition, the freshwater consumption rate was 39.734 tonne/h because the upper limit of one or more pipeline capacities was reached. The corresponding minimum upper limit of the freshwater supply rate was then determined to be 36.62 tonne/h with the proposed NLP-SC model. Thus, the overdesign level of the freshwater supply system in design 6.5.I should be at least 38.25%. Finally, note that a summary of the assessment findings is presented in Table 6.12.

3. Let us next evaluate the outcomes of implementing various revamp measures to design 6.5.II:

   a. The upper limit of the freshwater supply rate was first raised to 20 tonne/h. By solving the NLP-FI model, the FI_s can be improved slightly to 0.190. The reason for such a minor improvement is that the critical freshwater usage is 12.576 tonne/h, which is the result

TABLE 6.12

Comparison of Optimal Solutions of Utility Models Based on Design 6.5.I

Step	Model	Overdesign		Flexibility	Freshwater
		Freshwater	Pipelines	Index	Usage(ton/hr)
1	NLP-FI	30%	50%	0.765	26.489
2	NLP-FI	Relaxed	50%	1.351	39.734
2	NLP-SC	38.25%	50%	1	36.62

Source: Reprinted with permission from Li and Chang 2011, 3763–3774. Copyright 2011 American Chemical Society.

of one or more upper limits imposed upon pipeline capacity. Next, the model constraints were further relaxed by removing all capacity limits on the water flows in existing pipelines, and the upper limit of the freshwater supply rate was increased to 50 tonne/h (denoted as design 6.5.II-A). The resulting flexibility index has also been computed with the NLP-FI model, and this value is 0.398. The obtained critical conditions are presented in Table 6.13 and Figure 6.11. Note that the *dashed* line in Figure 6.11 means that there is no flow under the critical condition. These results also show that raising the freshwater capacity to any level higher than 42.01 tonne/h is useless.

b. It can be observed from Table 6.13 that the upper limits of C_{d1}^{in}, C_{u1}^{out}, C_{u2}^{out}, C_{u3}^{out}, FT_{t1}, and FT_{t2} are reached, and thus the corresponding inequalities are the active constraints. Obviously, the flexibility index can be increased only if the active constraints are relaxed. Thus, new auxiliary pipelines may be added so as to facilitate the relaxation of such constraints. Because it is clearly not feasible to lower the throughput of any wastewater treatment unit by feeding an extra water flow, only the possibilities of relaxing the first four inequality constraints are considered here. These considerations are summarized as follows:

i. Because sink $d1$ already accepts water streams from $t1$ and $t2$ in the present network, it is only necessary to consider other sources. Notice that the concentration at the sink (C_{d1}^{in}) is less likely to be lowered by adding a pipeline from any of the water-using units, that is, $u1$, $u2$, or $u3$ to sink $d1$. This is because the outlet concentrations of these units reach their maxima in the optimal solution, which is much larger than the allowed maximum value of C_{d1}^{in}. It should also be noted that dilution of effluent to sink $d1$ directly with freshwater $w1$ is not allowed in the present study. Therefore, it can be concluded that the active constraint corresponding to C_{d1}^{in} cannot be relaxed by introducing new pipelines.

ii. The option of adding an extra water flow to lower C_{u1}^{out} should be ignored because unit $u1$ has the lowest concentration limits on both inlet and outlet.

TABLE 6.13
Critical Operating Conditions of All Units in Design 6.5.II-A

Unit	$u1$	$u2$	$u3$	$t1$	$t2$	$d1$
Flow rate (ton/hr)	42.007	26.758	26.234	125.000	135.000	72.007
C^{in} (ppm)	0.100	18.241	18.241	164.724	5.561	10.000
C^{out} (ppm)	101.000	240.000	200.000	18.241	1.165	

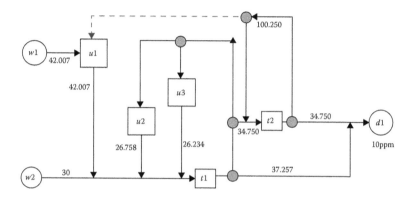

FIGURE 6.11 Critical operating conditions of design 6.5.II-A. (Reprinted with permission from Li and Chang 2011, 3763–3774. Copyright 2011 American Chemical Society.)

iii. Based on the optimality conditions of the water utilization system (Savelski and Bagajewicz, 2000), the used water from $u1$ can be partly reused in $u2$ to reduce C_{u2}^{out}. This mainly is because the allowed maximum outlet concentration of $u1$ is much less than that of $u2$, and there is still room for the inlet concentration of $u2$ to increase.

iv. For the same reason, pipelines from $u1$ to $u3$ and from $u3$ to $u2$ may be added to design 6.5.II-A.

The revised network is referred to as design 6.5.II-B, and the corresponding flexibility index found by solving the NLP-FI model is 1.4535. The resulting critical operating conditions are given in Table 6.14 and Figure 6.12. Notice that the added auxiliary pipelines are marked with blue *dotted* lines.

c. Design 6.5.II-B can then be reconfigured with the MINLP-NR model. By setting the cost coefficients C_{pl} and C_{w_1} to be 1 and 0.5, respectively, design 6.5.II-C can be obtained. The optimal solution is shown in Table 6.15 and Figure 6.13. In this case, the minimum total annual cost is 26.840, and the freshwater consumption rate is 27.680 tonne/h. Notice that one of the added pipelines is eliminated in the optimal network (as shown with a blue dashed line) and one existing branch is also removed (as shown with a black dashed line).

d. It should be noted that the optimal solution is not unique. Other alternatives can be easily created by slightly changing the initial

TABLE 6.14

Critical Operating Conditions of All Units in Design 6.5.II-B

Unit	u1	u2	u3	t1	t2	d1
Flow rate (ton/hr)	66.652	42.631	36.541	125.000	135.000	80
C^{in} (ppm)	1.000	80	50	185.00	15.762	10.000
C^{out} (ppm)	74.097	240.000	200.000	25.760	3.702	

Source: Reprinted with permission from Li and Chang 2011, 3763–3774. Copyright 2011 American Chemical Society.

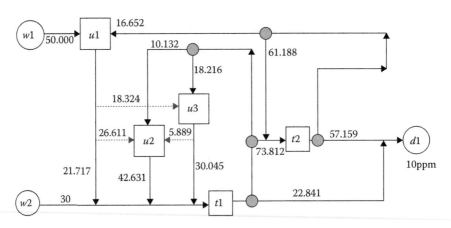

FIGURE 6.12 Critical operating conditions of design 6.5.II-B. (Reprinted with permission from Li and Chang 2011, 3763–3774. Copyright 2011 American Chemical Society.)

TABLE 6.15

Critical Operating Conditions of All Units in Design 6.5.II-C

Unit	u1	u2	u3	t1	t2	d1
Flow rate (ton/hr)	46.000	61.240	33.760	125.000	135.000	57.680
C^{in} (ppm)	1.000	68.261	46.712	178.578	10.535	10.000
C^{out} (ppm)	101.000	173.421	200.000	22.679	2.360	

Source: Reprinted with permission from Li and Chang 2011, 3763–3774. Copyright 2011 American Chemical Society.

guess or using a different solver. An example is given in Figure 6.14 (design 6.5.II-D). Although the objective value of this design is the same as that of Figure 6.13, only one of the three added auxiliary pipelines is kept in this solution.

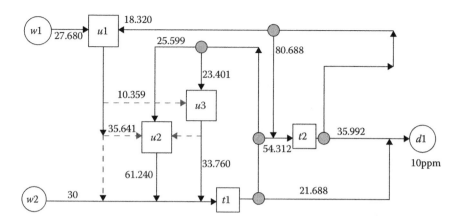

FIGURE 6.13 Critical operating conditions of design 6.5.II-C. (Reprinted with permission from Li and Chang 2011, 3763–3774. Copyright 2011 American Chemical Society.)

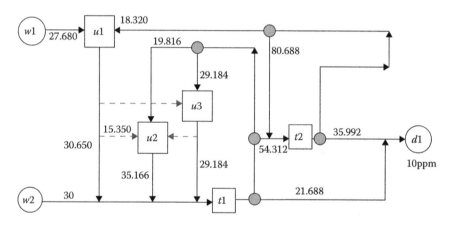

FIGURE 6.14 Alternative critical operating conditions of design 6.5.II-D. (Reprinted with permission from Li and Chang 2011, 3763–3774. Copyright 2011 American Chemical Society.)

6.6 MULTICONTAMINANT SYSTEMS

6.6.1 ITERATIVE EVALUATION OF THE FLEXIBILITY INDEX VIA SINGLE-VERTEX TESTS

Although the available solution strategies for evaluating the steady-state flexibility index have already been presented in Chapter 2, their basic model framework is still briefly repeated in this subsection for illustration clarity and completeness. As mentioned before, let us express the model constraints as

$$h_i(\mathbf{d}, \mathbf{z}, \mathbf{x}, \boldsymbol{\theta}) = 0 \quad \forall i \in \mathbb{I} \tag{6.56}$$

$$g_j(\mathbf{d}, \mathbf{z}, \mathbf{x}, \boldsymbol{\theta}) \leq 0 \quad \forall j \in \mathbb{J} \tag{6.57}$$

where

$$\mathbb{I} = \left\{ i \mid i \text{ is the label of an equality constraint} \right\}$$

$$\mathbb{J} = \left\{ j \mid j \text{ is the label of an inequality constraint} \right\}$$

Also in these constraints, $\mathbf{d}$ represents a vector in which all binary parameters in Equation 6.1 are stored; $\mathbf{z}$ denotes the vector of adjustable control variables; $\mathbf{x}$ is the vector of state variables; and θ denotes the vector of uncertain parameters (or multipliers), and these parameters are present in a space $\Gamma(\delta)$ defined as follows:

$$\Gamma(\delta) = \left\{ \theta^N - \delta \Delta \theta^- \le \theta \le \theta^N + \delta \Delta \theta^+ \right\} \tag{6.58}$$

where $\Delta\theta^+$ and $\Delta\theta^-$ denote the vectors of expected deviations in the positive and negative directions, respectively; $\delta \ge 0$ is a scalar variable.

The steady-state flexibility index FI_s was traditionally regarded as the maximum value of δ that renders all points in $\Gamma(\delta)$ feasible. In the single-contaminant cases, FI_s can usually be determined with the *active set method* by solving a nonconvex MINLP model (Riyanto and Chang, 2010). Although this approach is theoretically sound, there are a number of drawbacks for the multicontaminant applications. In particular, because of the need to invoke KKT conditions, it is often tedious to construct the corresponding MINLP model even for a moderately complex water network. Another more serious disadvantage can be attributed to the fact that the convergence of the optimization run cannot be guaranteed. This feature is especially unacceptable when, for the purpose of identifying the best revamp design in an evolutionary procedure, the model must be solved repeatedly for various combinations of the binary parameters in $\mathbf{d}$.

To overcome the aforementioned computational difficulties, the flexibility index is computed in the present case by solving the *flexibility test* problem iteratively according to the underlying principles of the vertex method. Specifically, for a given value of scalar variable δ and a given set of binary parameters $\mathbf{d}$, the feasibility of a water network design can be tested by carrying out the optimization run required by the following formulation:

$$\chi(\mathbf{d}) = \max_{k \in V} \min_{z,u} u$$

s.t.

$$h_i(\mathbf{d}, \mathbf{z}, \mathbf{x}, \theta^k) = 0 \quad \forall i \in \mathbb{I} \tag{6.59}$$

$$g_j(\mathbf{d}, \mathbf{z}, \mathbf{x}, \theta^k) \le u \quad \forall j \in \mathbb{J}$$

where V denotes the set of all vertices in $\Gamma(\delta)$ and θ^k is one of them in this set (i.e., vertex k). The design is considered feasible if $\chi(\mathbf{d}) \le 0$ and infeasible if otherwise.

It can also be observed from Equation 6.59 that the minimum values of u at all vertexes must be determined in this optimization problem. As mentioned before in the single-contaminant scenarios, the multicontaminant problems can also be simplified by checking only a *single* critical vertex. For the purpose of reducing computation load, the single-vertex test (i.e., Equation 6.59) is repeatedly performed to guide the search for determining the flexibility index FI_s. More specifically, a simple bisection search strategy is adopted to locate the maximum feasible δ by a set of given binary parameters in **d**. Following is a description of the proposed algorithm:

1. Let $n = 0$. Set the lower bound of the flexibility index to be $FI_n^{low} = 0$ and the upper bound FI_n^{up} an arbitrarily selected large number.

2. Let $\delta = \dfrac{FI_n^{up} + FI_n^{low}}{2}$ and apply the single-vertex flexibility test.

3. Let $n = n + 1$. If the test in step 2 is feasible, set $FI_n^{low} = \delta$ and $FI_n^{up} = FI_{n-1}^{up}$. Otherwise, set $FI_n^{low} = FI_{n-1}^{low}$ and $FI_n^{up} = \delta$.

4. Check if a given termination criterion (say, $FI_n^{up} - FI_n^{low} < \varepsilon$) is satisfied. If not, go to step 2. Otherwise, stop.

Note that an implied assumption in this procedure is that the test result for $\delta = FI_0^{up}$ is infeasible. If this is not the case with the selected initial guess, then the upper bound must be enlarged to satisfy this requirement.

Three examples are presented next to demonstrate the feasibility and superiority of the single-vertex search algorithm. All problems were solved on a PC that is equipped with an Intel® Core™2 Quad CPU Q9400 and 4.00 GB RAM (3.25 GB usable) 32-bit operating system platform. The single-vertex flexibility test model was coded with GAMS and solved with BARON, and the bisection search procedure was realized using MATLAB via the MATLAB-GAMS interface.

Example 6.6

Let us consider the nominal water network presented in Figure 6.15, in which one freshwater source (W_1), one secondary source (W_2), three water-using units (U_1, U_2, and U_3), two wastewater treatment units ($T1$ and $T2$), and a sink are involved. The model parameters for this example are presented in Table 6.16. Six uncertain multipliers are considered in this example, and their expected deviations are:

$$\Delta\theta_{C_{w2,A}}^{+} = \Delta\theta_{C_{w2,A}}^{-} = 0.1$$

$$\Delta\theta_{M_{U1,A}}^{+} = \Delta\theta_{M_{U1,A}}^{-} = \Delta\theta_{M_{U2,A}}^{+} = \Delta\theta_{M_{U2,A}}^{-} = \Delta\theta_{M_{U3,A}}^{+} = \Delta\theta_{M_{U3,A}}^{-} = 0.15$$

$$\Delta\theta_{RR1,A}^{+} = \Delta\theta_{RR1,A}^{-} = \Delta\theta_{RR2,A}^{+} = \Delta\theta_{RR2,A}^{-} = 0.03$$

Notice also that this example is taken from Riyanto and Chang (2010), and the flexibility index was found to be 0.32 with the active set method.

The convergence process of the bisection search with the single-vertex method is described in Figure 6.16. Notice that both the upper and lower bounds, if the flexibility indices are plotted at every iteration and their initial guesses, were set

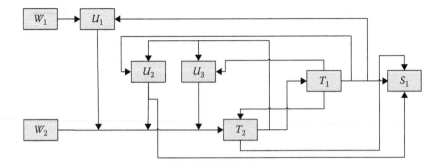

FIGURE 6.15   The nominal structure of the water network in Example 6.6. (Reprinted from *Chemical Engineering Science*, Vol. 102, Jiang, D., Chang, C.T., An algorithmic revamp strategy for improving operational flexibility of multi-contaminant water networks, 289–299. Copyright 2013 with permission from Elsevier.)

TABLE 6.16
The Model Parameters Used in Example 6.6

Parameter		Value	Parameter		Value
$F_{W_1}^{max}$	(tonne/hr)	30.0	$C_{u_3,A}^{out,max}$	(ppm)	200.0
F_{W_2}	(tonne/hr)	30.0	$F_{u_1}^{max}$	(tonne/hr)	125.0
$C_{W_1,A}$	(ppm)	0.1	$F_{u_2}^{max}$	(tonne/hr)	135.0
$\overline{C}_{W_2,A}$	(ppm)	150.0	$C_{s_1,A}^{max}$	(ppm)	10.0
$C_{u_1,A}^{in,max}$	(ppm)	1.0	$\overline{ML}_{u_1,A}$	(kg/hr)	4.0
$C_{u_1,A}^{out,max}$	(ppm)	101.0	$\overline{ML}_{u_2,A}$	(kg/hr)	5.6
$C_{u_2,A}^{in,max}$	(ppm)	80.0	$\overline{ML}_{u_3,A}$	(kg/hr)	4.5
$C_{u_2,A}^{out,max}$	(ppm)	240.0	$\overline{R}_{t_1,A}$		0.9
$C_{u_3,A}^{in,max}$	(ppm)	50.0	$\overline{R}_{t_2,A}$		0.8

Source: Reprinted from *Chemical Engineering Science*, Vol. 102, Jiang, D., Chang, C.T., An algorithmic revamp strategy for improving operational flexibility of multi-contaminant water networks, 289–299. Copyright 2013 with permission from Elsevier.

to be 16 and 0, respectively. It is clear that the search converges after about 10 iterations to the correct value. The computation time in this case is 152 sec; however, a much longer 571 sec is needed if the traditional vertex method is used to perform the flexibility test.

Example 6.7

Because the single-vertex strategy has only been applied to the single-contaminant systems in Section 6.5, it is, of course, desirable to find out if the same approach is also effective in multicontaminant applications. For this purpose, let us consider the nominal water network presented in Figure 6.17 and the corresponding model

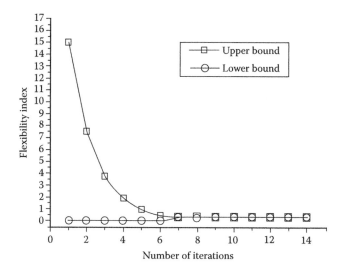

FIGURE 6.16 The convergence behavior of bisection searches in Example 6.6. (Reprinted from *Chemical Engineering Science*, Vol. 102, Jiang, D., Chang, C.T., An algorithmic revamp strategy for improving operational flexibility of multi-contaminant water networks, 289–299. Copyright 2013 with permission from Elsevier.)

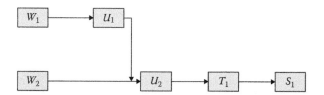

FIGURE 6.17 The nominal structure of the water network in Example 6.7. (Reprinted from *Chemical Engineering Science*, Vol. 102, Jiang, D., Chang, C.T., An algorithmic revamp strategy for improving operational flexibility of multi-contaminant water networks, 289–299. Copyright 2013 with permission from Elsevier.)

parameters in Table 6.17. Four uncertain multipliers are adopted in this example, and their expected deviations are

$$\Delta\theta^-_{Mlu_1,B} = \Delta\theta^-_{Mlu_2,B} = \Delta\theta^-_{Mlu_1,A} = \Delta\theta^-_{Mlu_2,A} = 0.1$$

$$\Delta\theta^+_{Mlu_1,A} = \Delta\theta^+_{Mlu_2,A} = 0.2$$

$$\Delta\theta^+_{Mlu_1,B} = \Delta\theta^+_{Mlu_2,B} = 0.3$$

The flexibility index of this network can be found to be 0.491 by the traditional active set method.

Although the critical vertex can be determined according to the selection criteria described in the previous section—the corner points that correspond to the

TABLE 6.17

The Model Parameters Used in Example 6.7

Parameter		Value	Parameter		Value
$F_{w_1}^{max}$	(tonne/hr)	35.0	$C_{u_1,B}^{in,max}$	(ppm)	15.0
F_{w_2}	(tonne/hr)	30.0	$C_{u_1,B}^{out,max}$	(ppm)	50.0
$F_{t_1}^{max}$	(tonne/hr)	125.0	$C_{u_2,B}^{in,max}$	(ppm)	60.0
$C_{w_1,A}$	(ppm)	0.1	$C_{u_2,B}^{out,max}$	(ppm)	90.0
$C_{w_1,B}$	(ppm)	1.0	$C_{t_1,B}^{in,max}$	(ppm)	90.0
$C_{w_2,A}$	(ppm)	100.0	$C_{s_1,B}^{max}$	(ppm)	30.0
$C_{w_2,B}$	(ppm)	10.0	$\overline{ML}_{u_1,A}$	(kg/hr)	2.0
$C_{u_1,A}^{in,max}$	(ppm)	1.0	$\overline{ML}_{u_2,A}$	(kg/hr)	5.0
$C_{u_1,A}^{out,max}$	(ppm)	101.0	$\overline{ML}_{u_1,B}$	(kg/hr)	1.0
$C_{u_2,A}^{in,max}$	(ppm)	80.0	$\overline{ML}_{u_2,B}$	(kg/hr)	2.0
$C_{u_2,A}^{out,max}$	(ppm)	240.0	$R_{t_1,A}$		0.9
$C_{t_1,A}^{in,max}$	(ppm)	185.0	$R_{t_1,B}$		0.6
$C_{s_1,A}^{max}$	(ppm)	30.0			

Source: Reprinted from *Chemical Engineering Science*, Vol. 102, Jiang, D., Chang, C.T., An algorithmic revamp strategy for improving operational flexibility of multi-contaminant water networks, 289–299. Copyright 2013 with permission from Elsevier.

upper bounds of all mass loads in the present example—each vertex has been tested in the proposed bisection search procedure to produce a corresponding "flexibility index." The results of 16 separate runs can be found in Table 6.18. It can be clearly observed that the correct FI_s value can indeed be obtained with the proposed single-vertex approach.

Example 6.8

This example in this section is adopted to demonstrate the advantage of the proposed computation strategy for solving large problems. Let us consider the complex nominal water network presented in Figure 6.18 in which three contaminants, one freshwater source (W_1), one secondary source (W_2), four water-using units (U_1, U_2, U_3, and U_4), a wastewater treatment unit (T_1), and a sink (S_1) are involved. The corresponding model parameters are presented in Table 6.19. It is also assumed that there are 18 uncertain multipliers, and the corresponding expected deviations are

$$\Delta\theta^+_{ML_{u_1},A} = \Delta\theta^-_{ML_{u_1},A} = \Delta\theta^+_{ML_{u_1},B} = \Delta\theta^-_{ML_{u_1},B} = \Delta\theta^+_{ML_{u_1},C} = \Delta\theta^-_{ML_{u_1},C} = 0.15$$

$$\Delta\theta^+_{ML_{u_2},A} = \Delta\theta^-_{ML_{u_2},A} = \Delta\theta^+_{ML_{u_2},B} = \Delta\theta^-_{ML_{u_2},B} = \Delta\theta^+_{ML_{u_2},C} = \Delta\theta^-_{ML_{u_2},C} = 0.25$$

$$\Delta\theta^+_{ML_{u_3},A} = \Delta\theta^-_{ML_{u_3},B} = \Delta\theta^+_{ML_{u_3},B} = \Delta\theta^-_{ML_{u_3},B} = \Delta\theta^+_{ML_{u_3},C} = \Delta\theta^-_{ML_{u_3},C} = 0.2$$

$$\Delta\theta^+_{ML_{u_4},A} = \Delta\theta^-_{ML_{u_4},A} = \Delta\theta^+_{ML_{u_4},B} = \Delta\theta^-_{ML_{u_4},B} = \Delta\theta^+_{ML_{u_4},C} = \Delta\theta^-_{ML_{u_4},C} = 0.3$$

$$\Delta\theta^+_{C_{w1,A}} = \Delta\theta^-_{C_{w1,A}} = \Delta\theta^+_{C_{w1,B}} = \Delta\theta^-_{C_{w1,B}} = \Delta\theta^+_{C_{w1,C}} = \Delta\theta^-_{C_{w1,C}} = 0.1$$

$$\Delta\theta^+_{C_{w2,A}} = \Delta\theta^+_{C_{w2,B}} = \Delta\theta^+_{C_{w2,C}} = 0.1$$

$$\Delta\theta^-_{C_{w2,A}} = \Delta\theta^-_{C_{w2,B}} = \Delta\theta^-_{C_{w2,C}} = 0.05$$

TABLE 6.18

The Flexibility Indices Obtained at All Vertices in Example 6.7

$\Delta\theta_{Mlu_1,A}$	$\Delta\theta_{Mlu_2,A}$	$\Delta\theta_{Mlu_1,B}$	$\Delta\theta_{Mlu_2,B}$	**Flexibility Index**
-	-	-	-	9.99
-	-	-	+	3.08
-	-	+	-	2.38
-	-	+	+	1.71
-	+	-	-	2.53
-	+	-	+	2.53
-	+	+	-	2.38
-	+	+	+	1.71
+	-	-	-	0.491
+	-	+	-	0.491
+	-	+	+	0.491
+	+	-	-	0.491
+	+	-	+	0.491
+	+	+	-	0.491
+	+	+	+	0.491

Source: Reprinted from *Chemical Engineering Science*, Vol. 102, Jiang, D., Chang, C.T., An algorithmic revamp strategy for improving operational flexibility of multi-contaminant water networks, 289–299. Copyright 2013 with permission from Elsevier.

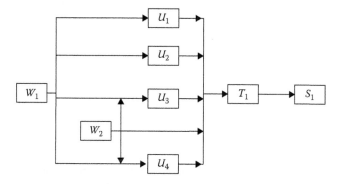

FIGURE 6.18 The nominal structure of the water network in Example 6.8. (Reprinted from *Chemical Engineering Science*, Vol. 102, Jiang, D., Chang, C.T., An algorithmic revamp strategy for improving operational flexibility of multi-contaminant water networks, 289–299. Copyright 2013 with permission from Elsevier.)

TABLE 6.19

The Model Parameters Used in Example 6.8

Parameter		Value	Parameter		Value
$F_{w_1}^{max}$	(tonne/hr)	25.0	$C_{t_1,B}^{in,max}$	(ppm)	250.0
F_{w_2}	(tonne/hr)	100.0	$C_{s_1,B}^{max}$	(ppm)	30.0
$F_{f_1}^{max}$	(tonne/hr)	150.0	$C_{u_1,C}^{in,max}$	(ppm)	10.0
$\bar{C}_{w_1,A}$	(ppm)	0.05	$C_{u_1,C}^{out,max}$	(ppm)	150.0
$\bar{C}_{w_1,B}$	(ppm)	1.0	$C_{u_2,C}^{in,max}$	(ppm)	100.0
$\bar{C}_{w_1,C}$	(ppm)	5.0	$C_{u_2,C}^{out,max}$	(ppm)	200.0
$\bar{C}_{w_2,A}$	(ppm)	100.0	$C_{u_3,C}^{in,max}$	(ppm)	100.0
$\bar{C}_{w_2,B}$	(ppm)	100.000	$C_{u_3,C}^{out,max}$	(ppm)	250.0
$\bar{C}_{w_2,C}$	(ppm)	50.0	$C_{u_4,C}^{in,max}$	(ppm)	100.0
$C_{u_1,A}^{in,max}$	(ppm)	1.0	$C_{u_4,C}^{out,max}$	(ppm)	250.0
$C_{u_1,A}^{out,max}$	(ppm)	50.0	$C_{t_1,C}^{in,max}$	(ppm)	250.0
$C_{u_2,A}^{in,max}$	(ppm)	50.0	$C_{s_1,C}^{max}$	(ppm)	70.0
$C_{u_2,A}^{out,max}$	(ppm)	250.0	$\overline{ML}_{u_1,A}$	(kg/hr)	0.1
$C_{u_3,A}^{in,max}$	(ppm)	100.0	$\overline{ML}_{u_2,A}$	(kg/hr)	2.0
$C_{u_3,A}^{out,max}$	(ppm)	200.0	$\overline{ML}_{u_3,A}$	(kg/hr)	5.0
$C_{u_4,A}^{in,max}$	(ppm)	100.0	$\overline{ML}_{u_4,A}$	(kg/hr)	7.0
$C_{u_4,A}^{out,max}$	(ppm)	200.0	$\overline{ML}_{u_1,B}$	(kg/hr)	0.3
$C_{t_1,A}^{in,max}$	(ppm)	200.0	$\overline{ML}_{u_2,B}$	(kg/hr)	2.0
$C_{s_1,A}^{max}$	(ppm)	50.0	$\overline{ML}_{u_3,B}$	(kg/hr)	3.0
$C_{u_1,B}^{in,max}$	(ppm)	5.0	$\overline{ML}_{u_4,B}$	(kg/hr)	6.0
$C_{u_1,B}^{out,max}$	(ppm)	120.0	$\overline{ML}_{u_1,C}$	(kg/hr)	0.4
$C_{u_2,B}^{in,max}$	(ppm)	70.0	$\overline{ML}_{u_2,C}$	(kg/hr)	2.0
$C_{u_2,B}^{out,max}$	(ppm)	170.0	$\overline{ML}_{u_3,C}$	(kg/hr)	3.0
$C_{u_3,B}^{in,max}$	(ppm)	120.0	$\overline{ML}_{u_4,C}$	(kg/hr)	6.0
$C_{u_3,B}^{out,max}$	(ppm)	200.0	$R_{t_1,A}$		0.8
$C_{u_4,B}^{in,max}$	(ppm)	120.0	$R_{t_1,B}$		0.9
$C_{u_4,B}^{out,max}$	(ppm)	200.0	$R_{t_1,C}$		0.6

Source: Reprinted from *Chemical Engineering Science*, Vol. 102, Jiang, D., Chang, C.T., An algorithmic revamp strategy for improving operational flexibility of multi-contaminant water networks, 289–299. Copyright 2013 with permission from Elsevier.

A problem of this scale cannot be solved with the active set method in a reasonable period (say, 24 hours). However, it took only 18 sec for the proposed search to converge, and a flexibility index of 0.0165 was found for the given system.

6.6.2 Evolutionary Identification of Revamp Designs

As mentioned previously, the ultimate objective of this work is to identify proper revamp designs for improving the operational flexibility of any given water network. The allowed revamp options are limited to those incorporated in the augmented superstructure (i.e., the new pipelines and/or treatment units). The design specifications of the embedded treatment units are assumed to be available in advance. Because the number of alternative structures increases exponentially with network complexity, a deterministic search strategy may fail to identify the optimal solution within a reasonable period. Therefore, a modified version of the genetic algorithm (GA) has been adopted to circumvent this drawback. Notice also that in a typical evolution procedure, every chromosome in a population can be expressed as a string of 0s and 1s, and this mechanism can be easily utilized for coding the structural optimization problem considered here.

In the proposed algorithm, the binary parameters defined in Equations 6.1 and 6.2—that is, the elements of a vector $\mathbf{d}$ in Equations 6.56 and 6.57—are encoded in every individual within a population. Essentially two alternative fitness measures (FM) can be considered for the purpose of generating revamp designs, that is,

$$\text{FM}_1 = FI \tag{6.60}$$

or

$$\text{FM}_2 = \frac{FI}{\sum d_{m,n}} \tag{6.61}$$

Note that the flexibility index FI_s in Equation 6.60 or 6.61 can be computed according to the single-vertex search algorithm described in Subsection 6.6.1. In particular, FM_1 is a measure of the operational flexibility of the revamped system, whereas FM_2 can be viewed as a cost-penalized version of FM_1.

The "fittest" individual(s) is obviously the one with the largest measure. The GA in MATLAB (MathWorks, 2016) is used to facilitate the necessary evolutionary computation procedure. Four standard evolutionary steps are performed for each generation: selection, recombination, mutation, and reinsertion.

In all cases presented in this chapter, the same GA parameters have been utilized in every run. A brief summary is given here:

- The population size was always set to be 100.
- The generation gap in the selection step was chosen to be 0.7.
- The crossover rate in the recombination step was 0.7.
- The mutation probability was fixed at the default value of $0.7/Lind$ in the mutation step, where $Lind$ is the chromosome length.

- In the reinsertion step, the offspring individuals were ranked according to their fitness measures, and the top 50% of them were selected to replace the same number of least-fit parents.

The evolutionary procedure was terminated if (1) the total number of evaluated generations exceeds 200 and (2) the largest fitness measure in a population stayed approximately the same for at least 30 generations.

Finally, because each FI_s is determined iteratively and this computation process can be very time consuming, an additional mechanism has been built into the MATLAB code to avoid repeating the same calculation for identical network configurations. In particular, the individuals and their corresponding fitness measures in the parent generation and those in all previous generations can be accumulated in a database. Every newly created individual in the offspring generation can be compared with the ones already stored there. If a match is identified, the corresponding fitness measure can be directly retrieved without the iterative computation.

The aforementioned evolution strategy has been tested extensively in the three examples summarized next.

Example 6.9

Let us consider the nominal water network in Figure 6.17 and the corresponding model parameters in Table 6.17. The uncertain parameters are the same as those described in Example 6.7. Without adding any more water treatment units, there are 11 new connections in the augmented superstructure. Additional treatment units could drastically increase the number of new connections in the superstructure. For example, this figure is raised from 11 to 32 if two new treatment units are allowed.

By using FM_1 as the fitness measure, more than one network structure was identified with the GA-based method. It was observed that two new connections, that is, (T_1, U_1) and (T_1, U_2), were embedded in all revamp options and, in fact, the highest FI_s value (= 2.6953) could also be achieved with these two indispensable additions only (see Figure 6.19). On the other hand, the second fitness measure was also adopted in an additional GA run to address the need to limit piping costs in revamp designs. In fact, exactly one new pipeline (T_1, U_2) was called for in the optimum solution obtained with FM_2 (see Figure 6.20). Notice that although the flexibility index of this structure was slightly decreased to 2.3828, the piping cost was obviously also lower than that of Figure 6.19. Finally, it was found that although adding the aforementioned new pipeline(s) is quite effective for flexibility enhancement, the given system could not be further improved with any new treatment unit.

Example 6.10

To show the potential benefits of installing additional treatment units, let us consider a single-contaminant system studied by Riyanto and Chang (2010). The nominal network structure of this problem is essentially the same as that presented

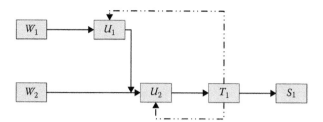

FIGURE 6.19 The revamp design obtained according to FM_1 in Example 6.9. (Reprinted from *Chemical Engineering Science*, Vol. 102, Jiang, D., Chang, C.T., An algorithmic revamp strategy for improving operational flexibility of multi-contaminant water networks, 289–299. Copyright 2013 with permission from Elsevier.)

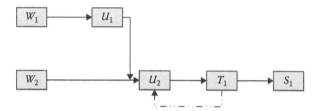

FIGURE 6.20 The revamp design obtained according to FM_2 in Example 6.9. (Reprinted from *Chemical Engineering Science*, Vol. 102, Jiang, D., Chang, C.T., An algorithmic revamp strategy for improving operational flexibility of multi-contaminant water networks, 289–299. Copyright 2013 with permission from Elsevier.)

in Figure 6.18, and the corresponding model parameters are given in Table 6.20. Seven uncertain multipliers were considered in their original work, and the following expected deviations were adopted:

$$\Delta\theta^+_{ML_{u1},A} = \Delta\theta^-_{ML_{u1},A} = \Delta\theta^+_{ML_{u2},A} = \Delta\theta^-_{ML_{u2},A} = 0.15$$

$$\Delta\theta^+_{ML_{u3},A} = \Delta\theta^-_{ML_{u3},A} = \Delta\theta^+_{ML_{u4},A} = \Delta\theta^-_{ML_{u4},A} = 0.15$$

$$\Delta\theta^-_{F^{max}_{w1}} = \Delta\theta^+_{F^{max}_{w1}} = 0.03$$

$$\Delta\theta^+_{C_{w1},A} = \Delta\theta^-_{C_{w1},A} = \Delta\theta^+_{C_{w2},A} = 0.1$$

$$\Delta\theta^-_{C_{w2},A} = 0.05$$

The flexibility index of this original network was found to be 0.249 with the active set method.

Without incorporating any additional treatment units, the number of new connections in the augmented superstructure can be found to be 25. The same FI_s value (i.e., 0.6445) was obtained by using either FM_1 or FM_2 in the GA evolution procedure. Three optimal structures were generated in the latter case

TABLE 6.20

The Model Parameters Used in Example 6.10

Parameter		Value	Parameter		Value
$\bar{F}_{w_1}^{max}$	(tonne/hr)	25.0	$C_{u_3,A}^{out,max}$	(ppm)	200.0
F_{w_2}	(tonne/hr)	100.0	$C_{u_4,A}^{in,max}$	(ppm)	100.0
F_n^{max}	(tonne/hr)	125.0	$C_{u_4,A}^{out,max}$	(ppm)	200.0
$\bar{C}_{w_1,A}$	(ppm)	0.05	$C_{t_1,A}^{in,max}$	(ppm)	200.0
$\bar{C}_{w_2,A}$	(ppm)	100.0	$C_{s_1,A}^{max}$	(ppm)	50.0
$C_{u_1,A}^{in,max}$	(ppm)	1.0	$\overline{ML}_{u_1,A}$	(kg/hr)	0.1
$C_{u_1,A}^{out,max}$	(ppm)	50.0	$\overline{ML}_{u_2,A}$	(kg/hr)	2.0
$C_{u_2,A}^{in,max}$	(ppm)	50.0	$\overline{ML}_{u_3,A}$	(kg/hr)	5.0
$C_{u_2,A}^{out,max}$	(ppm)	250.0	$\overline{ML}_{u_4,A}$	(kg/hr)	7.0
$C_{u_3,A}^{in,max}$	(ppm)	100.0	$R_{t_1,A}$		0.8

Source: Reprinted from *Chemical Engineering Science*, Vol. 102, Jiang, D., Chang, C.T., An algorithmic revamp strategy for improving operational flexibility of multi-contaminant water networks, 289–299. Copyright 2013 with permission from Elsevier.

(see Figure 6.21), and each contains two new connections. Specifically, the added pipelines in these three designs are (a) (U_4, S_1) and (T_1, U_2); (b) (U_4, S_1) and (T_1, U_3); (c) (U_4, S_1) and (T_1, U_4).

If one additional wastewater treatment unit (X_1) (with a removal ratio of 0.9) is allowed to be added to the existing water network, 39 new connections are present in the augmented superstructure. It is obviously impractical to evaluate all 2^{39} possible structures. By following the proposed GA evolution procedure with either FM₁ or FM₂ as the fitness measure, the maximum FI_s was raised to the same value of 6.660. Only one solution was produced by using the latter measure (see Figure 6.22), and this revamp design requires four new connections: (T_1, U_2), (U_4, X_1), (X_1, U_3), and (X_1, U_4). In this case, FM₂ = 6.660 / 4 = 1.665.

It should be noted that the highest FI_s value reported by Riyanto and Chang (2010) was only 1.604 for the present example. The corresponding revamp design consists of one new treatment unit (with a removal ratio of 0.9) and three new connections: (U_4, X_1), (X_1, W_2), and (X_1, S_1). For comparison purposes, let us also compute the second fitness measure for this design: FM₂ = 1.604 / 3 = 0.535. Thus, it can be observed from the values of both FI_s and FM₂ that the proposed programming-based revamp strategy clearly outperforms the heuristic approach in this example.

Example 6.11

In this last example, let us consider the nominal structure described in Example 6.8.
The first scenario is concerned with an augmented superstructure in which additional wastewater treatment units are *not* allowed. Thus, the total number of new connections should be 25. A maximum FI_s value of 1.331 was obtained

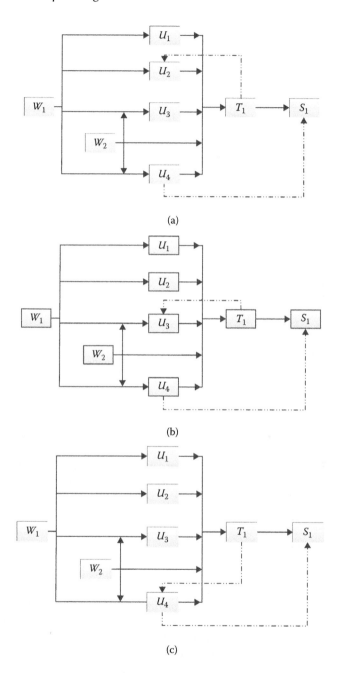

(a)

(b)

(c)

FIGURE 6.21 The revamp designs obtained according to FM_2 in Example 6.10 (without new treatment units): (a) structure 6.10.1; (b) structure 6.10.2; (c) structure 6.10.3. (Reprinted from *Chemical Engineering Science*, Vol. 102, Jiang, D., Chang, C.T., An algorithmic revamp strategy for improving operational flexibility of multi-contaminant water networks, 289–299. Copyright 2013 with permission from Elsevier.)

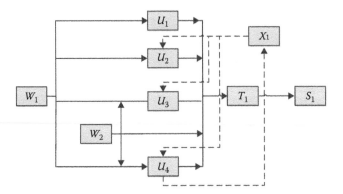

FIGURE 6.22 The revamp design (structure 6.10.4) obtained according to FM_2 in Example 6.10 (with one new treatment unit). (Reprinted from *Chemical Engineering Science*, Vol. 102, Jiang, D., Chang, C.T., An algorithmic revamp strategy for improving operational flexibility of multi-contaminant water networks, 289–299. Copyright 2013 with permission from Elsevier.)

by using FM_1 as the fitness measure. On the other hand, the flexibility index was reduced to 0.859 with the second fitness measure FM_2. Two alternative structures were obtained (see Figure 6.23), and only one new pipeline was needed in each design, that is, (a) (T_4, U_1) and (b) (U_1, U_4).

If two additional treatment units (with the same removal ratio of 0.9 for all contaminants) are allowed in the augmented superstructure, the total number of new connections should be increased to 53. Again FM_1 and FM_2 were used as the fitness measures in two separate GA runs. The resulting FI values were determined to be 3.332 and 3.327, respectively. The required computation time for the former run was 15,385 sec, whereas that for the latter was 13,920 sec. Finally, it was found that by maximizing the second fitness measure, one new treatment unit (X_2) and three new pipelines, that is, (U_4, X_2), (X_2, U_2) and (X_2, U_4), were selected in the optimal revamp design (see Figure 6.24).

6.7 CONCLUDING REMARKS

1. A novel heuristic strategy is developed in this chapter to improve the operation resiliency of any existing water network by relaxing the active constraints identified in the optimal solution of the flexibility index model. Each of the proposed structural modifications—that is, introducing the auxiliary pipelines, upgrading the existing treatment units, and installing new treatment units—may be used for this purpose when increasing the upper limit of the freshwater supply rate is not effective or not possible. The appropriate revamp options can be selected systematically with the aid of proposed design heuristics. From the results obtained so far in case studies, it can be concluded that this simple heuristic approach can provide good starting points for a rigorous method and reasonably good designs in practical applications.

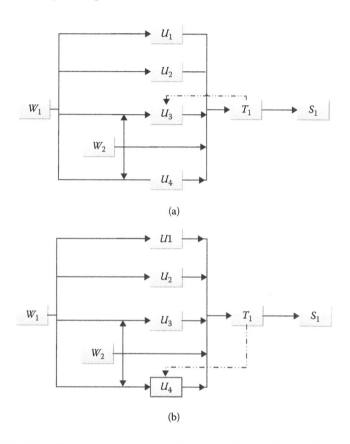

(a)

(b)

FIGURE 6.23 The revamp designs obtained according to FM$_2$ in Example 6.11 (without new treatment units): (a) structure 6.11.1; (b) structure 6.11.2. (Reprinted from *Chemical Engineering Science*, Vol. 102, Jiang, D., Chang, C.T., An algorithmic revamp strategy for improving operational flexibility of multi-contaminant water networks, 289–299. Copyright 2013 with permission from Elsevier.)

2. A programming approach is also presented in this chapter to assess the operational flexibility of given water networks. The flexibility of a given water network can be improved by relaxing the upper limit of the freshwater supply rate and/or incorporating structural modifications. It has been shown in the case studies that the proposed assessment procedure is feasible and efficient. Furthermore, the following conclusions can also be drawn from the optimization results obtained in the examples: (a) The proposed NLP-FI model is much easier to solve than the existing active constraint-based formulation, and the same quality solutions can be obtained in both cases. (b) The traditional ad hoc approach to set the overdesign levels on the freshwater supply system and pipelines may not be sufficient to overcome all uncertain disturbances. The proposed NLP-SC model represents a better alternative, which could be used to exactly determine the minimum

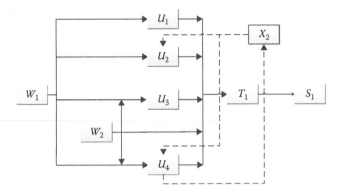

FIGURE 6.24 The revamp design (structure 6.11.3) obtained according to FM_2 in Example 6.11 (with one new treatment unit). (Reprinted from *Chemical Engineering Science*, Vol. 102, Jiang, D., Chang, C.T., An algorithmic revamp strategy for improving operational flexibility of multi-contaminant water networks, 289–299. Copyright 2013 with permission from Elsevier.)

freshwater supply capacity. (c) The proposed MINLP-NR model can be used to add/remove pipelines automatically so as to achieve the desired level of operational flexibility.

3. A programming-based approach has been developed to revamp any given water network for the purpose of flexibility enhancement. To alleviate the overwhelming manual and computational efforts required in deriving and solving the conventional flexibility index model with the active set method, a simple strategy is devised in this study to determine FI by repeatedly performing the flexibility test in a bisection search procedure. By incorporating this solution technique in a GA, more flexible revamp designs can be identified automatically by two alternative fitness measures. A series of numerical experiments and case studies have been carried out in this work to verify the feasibility and effectiveness of the proposed approach. In every example studied so far, the converged optimization results were not only satisfactory, but also were obtained within a reasonable period.

REFERENCES

Al-Redhwan, S.A., Crittender, B.D., Lababidi, H.M.S., 2005. Wastewater minimization under uncertain operational conditions. *Computers & Chemical Engineering* 29, 1009–1021.

Alva-Argaez, A., Kokossis, A.C., Smith, R., 1998. Wastewater minimisation of industrial systems using an integrated approach. *Computers & Chemical Engineering* 22, S741–S744.

Balakrishna, S., Biegler, L.T., 1992. Targeting strategies for the synthesis and energy integration of nonisothermal reactor networks. *Industrial & Engineering Chemistry Research* 31, 2152–2164.

Byers, W., Lindgren, G., Noling, C., Peters, D., 2003. *Industrial Water Management: A Systems Approach*. Center for Waste Reduction Technologies, American Institute of Chemical Engineers, New York, Wiley.

Chang, C.T., Li, B.H., Liou, C.W., 2009. Development of a generalized mixed integer nonlinear programming model for assessing and improving the operational flexibility of water network designs. *Industrial & Engineering Chemistry Research* 48, 3496–3504.

Feng, X., Seider, W.D., 2001. New structure and design methodology for water networks. *Industrial & Engineering Chemistry Research* 40, 6140–6146.

Huang, C.H., Chang, C.T., Ling, H.C., Chang, C.C., 1999. A mathematical programming model for water usage and treatment network design. *Industrial & Engineering Chemistry Research* 38, 2666–2679.

Jiang, D., Chang, C.T., 2013. An algorithmic revamp strategy for improving operational flexibility of multi-contaminant water networks. *Chemical Engineering Science* 102, 289–299.

Karuppiah, R., Grossmann, I.E., 2006. Global optimization for the synthesis of integrated water systems in chemical processes. *Computers & Chemical Engineering* 30, 650–673.

Karuppiah, R., Grossmann, I.E., 2008. Global optimization of multiscenario mixed integer nonlinear programming models arising in the synthesis of integrated water networks under uncertainty. *Computers & Chemical Engineering* 32, 145–160.

Li, B.H., Chang, C.T., 2011. Efficient flexibility assessment procedure for water network designs. *Industrial & Engineering Chemistry Research* 50, 3763–3774.

MathWorks, 2016. *Genetic algorithm*. The MathWorks, Inc., Massachusetts.

Raspanti, C.G., Bandoni, J.A., Biegler, L.T., 2000. New strategies for flexibility analysis and design under uncertainty. *Computers & Chemical Engineering* 24, 2193–2209.

Riyanto, E., 2009. A heuristical revamp strategy to improve operational flexibility of existing water networks, Department of Chemical Engineering. National Cheng Kung University, Tainan.

Riyanto, E., Chang, C.T., 2010. A heuristic revamp strategy to improve operational flexibility of water networks based on active constraints. *Chemical Engineering Science* 65, 2758–2770.

Savelski, M.J., Bagajewicz, M.J., 2000. On the optimality conditions of water utilization systems in process plants with single contaminants. *Chemical Engineering Science* 55, 5035–5048.

Bagajewicz, M.J., Rivas, M., and Savelski, M.J., 1999. *A New Approach to the Design of Water Utilization Systems with Multiple Contaminants in Process Plants*. Presented at the 1999 AICHE National Meeting, Dallas.

Takama, N., Kuriyama, T., Shiroko, K., Umeda, T., 1980. Optimal water allocation in a petroleum refinery. *Computers & Chemical Engineering* 4, 251–258.

Tan, R.R., Cruz, D.E., 2004. Synthesis of robust water reuse networks for single-component retrofit problems using symmetric fuzzy linear programming. *Computers & Chemical Engineering* 28, 2547–2551.

Tan, R.R., Foo, D.C.Y., Manan, Z.A., 2007. Assessing the sensitivity of water networks to noisy mass loads using Monte Carlo simulation. *Computers & Chemical Engineering* 31, 1355–1363.

Wang, Y.P., Smith, R., 1995. Wastewater minimization with flowrate constraints. *Chemical Engineering Research & Design* 73, 889–904.

Zhang, Z., Feng, X., Qian, F., 2009. Studies on resilience of water networks. *Chemical Engineering Journal* 147, 117–121.

7 Steady-State and Volumetric Flexibility Analyses for Membrane Modules and Heat Exchanger Networks

Although the steady-state flexibility analysis has been applied successfully in the previous chapter for revamping water networks, it should be noted that FI_s may not always be a representative performance measure in practical applications. In some cases, it is beneficial to also consider the volumetric flexibility index (FI_v) as an alternative metric. To demonstrate the merits of this multicriteria approach, both flexibility analyses are applied in this chapter to the designs of two different types of realistic systems: the membrane module and the heat exchanger network (HEN). The detailed discussions are presented one at a time in the sequel.

7.1 MEMBRANE MODULES

7.1.1 Background

Separation processes play a remarkable role in the chemical and pharmaceutical industries, where they account for 40%–70% of both capital and operating costs (Adler et al., 2000). Starting in the late 1960s, membrane processes have gradually attracted interest for industrial applications and have provided feasible alternatives for, and have also been combined with, more traditional purification and separation processes (such as distillation, evaporation, adsorption, extraction, chromatography, etc.). This has been motivated by the benefits that membrane technology can offer over conventional techniques in terms of economy, environment, and safety (Geens et al., 2007; Lin and Livingston, 2007; Vandezande et al., 2008). Particularly in liquid processing, membrane operations may be classified into three simple operating modes: concentration, solvent exchange, and purification.

Although membrane filtration is, in principle, very promising, its widespread industrial use has not been realized due to material- and process-focused challenges. Numerous studies on the modeling, design, and optimization of membrane filtration systems have already been performed in recent years, for example, see Kim et al. (2014), Lin and Livingston (2007), and Siew et al. (2013a, b). Lin and

Livingston (2007) investigated multistage, continuous, countercurrent membrane cascades with recycling of the retentate for the solvent exchange of methanol and toluene. Apart from the experimental studies concerning the effects of cascade parameters on the separation, they performed numerical simulations based on a shortcut model and compared various cascade setups. In the recent works of Siew and his coworkers, organic solvent nanofitration (OSN) membrane cascades were applied for American Petroleum Institute (API) solute fractionation and concentration (Siew et al., 2013a, b). In their experimental study, they demonstrated that the separation performance of a stripping membrane cascade for solute fractionation was dependent on the relative permeability of the solutes through the membranes. They showed experimentally that a three-stage stripping cascade configuration leads to a significant reduction of the solvent use while maintaining high API purities (Siew et al., 2013a). In their subsequent paper, they demonstrated that membrane cascades can lead to sufficient API rejections, and they also performed model validation with experimental data and analyzed the effects of operating conditions (e.g., reflux ratios) on separation efficiencies with a McCabe-Thiele representation of the concentrations along the membrane cascade (Siew et al., 2013b). Kim et al. (2014) successfully applied a two-stage membrane cascade for constant volume diafiltration of PEG-400 and PEG-2000 in acetonitrile. Although successful applications were reported, it should be noted that the aforementioned works focused only upon experimental and simulation verification, whereas the important issues concerning operational flexibility have never been addressed.

Dealing with uncertainties is one of the practical issues that must be addressed in designing and operating any separation process. A realistic membrane module design is expected to be fully functional in the presence of uncertain operating temperature, feed flow rate, and concentration. As mentioned previously in Chapter 2, the ability of a system to maintain feasible operation despite unexpected disturbances is referred to as its operational flexibility. Grossmann and his coworkers first proposed a formal definition of operational feasibility/flexibility and developed a quantitative performance measure accordingly to facilitate performance evaluation (Grossmann and Floudas, 1987; Swaney and Grossmann, 1985a, b). More specifically, their steady-state flexibility index (denoted in this book as FI_s) was computed numerically by solving a multilevel optimization problem. On the other hand, Lai and Hui (2008) suggested using an alternative metric—the volumetric flexibility index (denoted in this book as FI_v)—to complement the conventional steady-state approach. Essentially, this index can be viewed in 3-D as the volumetric fraction of the feasible region inside a cube bounded by the expected upper and lower limits of uncertain process parameters. Because the total volume of the feasible region is calculated without the need to specify a nominal point and/or to identify the biggest inscribable cube in the feasible region, the magnitude of FI_v may be more closely linked to process flexibility in cases when the feasible regions are nonconvex. Adi et al. (2016) developed a computation method to accurately quantify the value of FI_v in a given system based on the Delaunay triangulation technique. Finally, it should be noted the numerical procedures for computing FI_s and FI_v have already been detailed in Chapters 2 and 3, respectively.

7.1.2 MODEL FORMULATIONS

A membrane module can be described with a simple mathematical model according to Figure 7.1. The corresponding material balance constraints can be imposed upon the total and component flow rates of feed and product streams. For a system with X components, the material balances can be written as

$$F = R + P \tag{7.1}$$

$$cF_x F = cR_x R + cP_x P \tag{7.2}$$

$$sc = \frac{P}{F} \tag{7.3}$$

$$sc^l \leq sc \tag{7.4}$$

$$sc^h \geq sc \tag{7.5}$$

$$cP_{x,\min} \leq cP_x \tag{7.6}$$

where $x \in \{1, 2, \cdots, X\}$; F is the total flow rate of fresh feed to the membrane module; cF_x is the weight fraction of component x in fresh feed; R is the total flow rate of retentate; cR_x is the weight fraction of component x in retentate; P is the total flow rate of permeate; cP_x is the weight fraction of component x in permeate; sc is the stage cut of the membrane module (i.e., the ratio between the permeate flow rate drawn from the module and the feed flow rate); sc^l and sc^h, respectively, denote the lower and upper bounds of stage cut; and $cP_{x,\min}$ is the minimum purity requirement in the permeate. Clearly the following constraints must also be included in the following mathematical model:

$$\sum_{x=1}^{X} cF_x = 1 \tag{7.7}$$

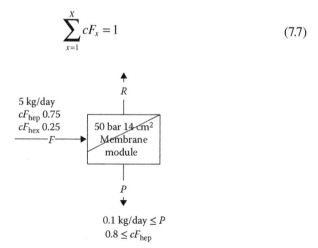

FIGURE 7.1 A typical membrane module and its operational constraints.

$$\sum_{x=1}^{X} cR_x = 1 \tag{7.8}$$

$$\sum_{x=1}^{X} cP_x = 1 \tag{7.9}$$

Although various different models are available for characterizing membrane performance, the solution-diffusion model for the OSN membrane is used in this work for illustration simplicity. In a solution-diffusion model, solute transport is expressed as a function of certain physicochemical and structural parameters (Marchetti and Livingston, 2015), and the implied assumption is that permeate is first adsorbed onto membrane and then diffuses along the concentration gradient (Wijmans and Baker, 1995). Separation is due to different rates of sorption and diffusion among the solutes. According to Marchetti and Livingston (2015), the embedded transport phenomena can be reliably described with this solution-diffusion model. Moreover, because such a model may be used to characterize the fluxes of both solute and solvent through the membranes, the model formulation of multicomponent separation can be greatly simplified, and thus, this flux model can be easily integrated into the unit models (Abejon et al., 2014). The following is also assumed:

- The axial pressure drop (i.e., the pressure drop across the membrane) is negligible compared to the actual operating pressure.
- The concentrations, temperature, and pressure in both compartments of a membrane module are homogeneous, so the membrane module can be modeled as two lumped systems.
- The heat generated by irreversible thermodynamic processes is negligible, so each membrane stage operates isothermally.

The corresponding transport model can therefore be written as

$$P = AJ\Delta P \tag{7.10}$$

$$J = \sum_{x=1}^{X} j_x \tag{7.11}$$

$$j_x = p_x \left(cR_x - \frac{\gamma_{P,x}}{\gamma_{R,x}} cP_x e^{\frac{v_x \cdot \Delta P}{R \cdot T}} \right) \tag{7.12}$$

$$\gamma_x = \sum_{i=0}^{n} a_i c_x^i \tag{7.13}$$

$$cP_xJ = j_x \tag{7.14}$$

$$P^l \leq P \tag{7.15}$$

$$P^h \geq P \tag{7.16}$$

where A is the effective area; ΔP is the pressure drop applied to the membrane module; J is the total flux through membrane; j_x is the flux of component x; p_x is the permeability coefficient of component x; v_x is the molar volume of component x; R is the ideal gas constant; T is the operating temperature; γ_x is the activity coefficient derived from the UNIFAC method, which could be approximated with polynomials; and P^l and P^h denote the lower and upper bounds of permeate flow rate, respectively. Note that Equation 7.12 is the primary nonlinear function in the mathematical model if the activity coefficient γ_x is assumed to be constant or linear; otherwise, Equation 7.13 will be the other nonlinear function. The effective area (A), pressure drop (ΔP), temperature (T), and stage cut (sc) should be considered as the design parameters in flexibility analyses, whereas the total and component fluxes (J and j_x) and the component concentrations in retentate and permeate streams (cR_x and cP_x) should be the state variables. In the present model, there are no control variables, and the uncertain parameters should be the total flow rate and the component concentration in feed (i.e., F and cF_x).

7.1.3 CASE STUDIES

In the following case studies, the operational flexibility of a lab-scale OSN membrane module is evaluated according to two alternative metrics: FI_s and FI_v. The flexibility analyses can be carried out to determine whether the membrane module is operable when the feed flow rate and concentration are expected to fluctuate within $\pm30\%$ from the nominal value. More specifically, let us introduce two uncertain multipliers (θ_F and θ_{cF}) as follows:

$$F = \theta_F F^N \tag{7.17}$$

$$cF_{heptane} = \theta_{cF} cF_{heptane}^N \tag{7.18}$$

where $1 - 0.3 \leq \theta_F, \theta_{cF} \leq 1 + 0.3$. The membrane module under consideration is designed for purifying a binary mixture of heptane and hexadecane, with a 75/25 weight fraction, that is, $cF_{heptane}^N = 0.75$, and the membrane is operated at 50 bar pressure to process 5 kg/day of feed, that is, $F^N = 5$. It is expected that the heptane purity in the permeate stream can reach no less than 80 wt. % by using a module with effective area of 14 cm^2. In the first scenario considered here, the operating conditions of the membrane module are constrained so that (1) the stage cut of the membrane is in the range of 0.3 to 0.7 and (2) the membrane module operates at room temperature (i.e., 25°C). The molar volumes of heptane and hexadecane can be found in the literature to be 1.482×10^{-4} m^3 / mol and 2.997×10^{-4} m^3 / mol, respectively. The permeability coefficients of both components for the particular module under consideration are $6.24 \ \text{kg} / \text{bar} \cdot \text{m}^2 \cdot \text{h}$ and $0.95 \ \text{kg} / \text{bar} \cdot \text{m}^2 \cdot \text{h}$. Because the OSN membranes at hand reject more hexadecane than heptane, the retentate stream should contain a higher mass percentage of hexadecane and, conversely, the permeate stream more heptane.

In Case 1 of the present example, FI_s can be found to be 0.003, with the active constraint located at the minimum stage cut of 0.3. The hypercube volume bounded by the expected upper and lower limits of uncertain multipliers is 0.36 units, and the corresponding volume of the feasible region is 0.16 units. Hence, it can be easily calculated that $FI_v = \dfrac{0.16}{0.36} = 0.44$. It can also be concluded that although a very pessimistic assessment is revealed by FI_s, the system actually can still be operable if the uncertain multiplier of feed concentration θ_{cF} is in the positive region (see Figure 7.2a). It also can be observed that the minimum stage cut constraint is shown at the bottom right of the feasible region, and the minimum required heptane purity constraint of 80% weight fraction is shown at the bottom left of the feasible region.

It should also be noted that in the aforementioned membrane model, the only constraint that could be compromised may be the one associated with the minimum stage cut. Thus, in the second case, the lower bound of the stage cut is relaxed from 0.3 to 0.2 (which means that a lower permeate flow rate is allowed in the operation). As a result, FI_s can be increased to 0.02, and the corresponding active constraint is associated with the minimum heptane purity. Figure 7.2b shows the relaxed feasible region where the minimum stage cut constraint is no longer affecting the feasible region. The resulting FI_s is 0.53 $(= \dfrac{0.19}{0.36})$ in this case.

One may wonder if the operating temperature can also be treated as an uncertain parameter because the ambient conditions are clearly uncontrollable. Thus, in Case 3 of this example, Case 1 is repeated under the assumption that T in Equation 7.12 may also vary ±30% from its nominal value (25°C). Consequently, the feasible region is now three-dimensional because there are three uncertain multipliers corresponding to the operating temperature, feed flow rate, and concentration, respectively. In this case FI_s is still found to be 0.003, and the corresponding active constraint is at the minimum stage cut of 0.3. The volume of a hypercube bounded by the expected upper and lower limits of uncertain parameters is 0.216 units, and the volume of the feasible region is now 0.0934 units. Hence, $FI_v = \dfrac{0.0934}{0.216} = 0.432$. From Figure 7.3, it can be observed that the impacts of temperature variation are quite linear on the feasible region and thus do not change the system flexibility significantly. One would therefore conclude that temperature uncertainty exerts little or no influence and the membrane system can be operated in a wide temperature range. A similar behavior may also be expected for Case 2.

From Case 1 and Case 2, it can also be observed that the minimum heptane purity of 80 wt. % is difficult to achieve when the feed concentration is at its lower

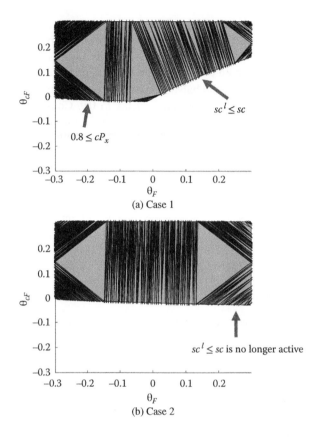

FIGURE 7.2   Feasible regions in (a) Case 1 and (b) Case 2 of the membrane module.

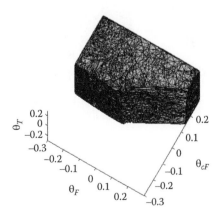

FIGURE 7.3   Feasible region in Case 3 for the membrane module.

bound. To increase the capability of the membrane module for the purifying feed, it may be necessary to select a new membrane with better separation character-istics and with a wider range of stage cut ratios. In Case 4, it is assumed that the permeability coefficients of both components in the new membrane module can be doubled to 12.48 kg / bar · m^2 · h and halved to 0.47 kg / bar · m^2 · h, respectively. The corresponding stage cut is now in the range of 0.1 to 0.9. Consequently, FI_s can be increased to 0.1 and the active constraint located at the lower bound of heptane purity. Figure 7.4 shows the corresponding feasible region where the minimum stage cut constraint is no longer a dominant factor. Instead, the maximum stage cut con-straint now appears to be active. In addition, $FI_v = \dfrac{0.24}{0.36} = 0.67$. Notice also that a more than 50% increase in FI_v (i.e., from 0.44 to 0.67) can be achieved with this new membrane module. Based on this observation, one would anticipate the proposed new membrane module should be more fault tolerant than the ones discussed previ-ously in Case 1 and Case 2.

7.1.4 CONCLUDING REMARKS

The steady-state and volumetric flexibility indices have been computed for the membrane modules discussed in four case studies. The feasible regions for oper-ating the membrane modules can be efficiently identified and the critical design constraints can be analyzed, respectively, to provide insights for identifying flex-ible designs. Based on the flexibility indices obtained in these studies, one can see that the upstream disturbances in feed flow rate and concentration exert profound impacts on operability, whereas the operating temperature is relatively unim-portant. A wider stage cut range and/or a higher separation factor is expected to increase operational flexibility. To facilitate comprehensive assessment, further

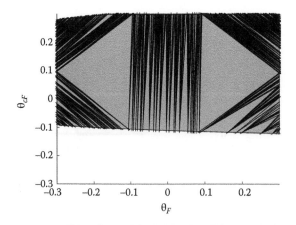

FIGURE 7.4   Feasible region in Case 4 for the membrane module.

works should be carried out for membrane cascades and with more rigorous transport models.

7.2 HEAT EXCHANGER NETWORKS

7.2.1 BACKGROUND

The HEN is an indispensable component in almost any chemical process. From the perspective of energy flows, three interactive components—the process, the utility system, and the HEN—must all be properly tied together and coordinated to form an operable plant (Aaltola, 2002). A large portion of the total annual cost (TAC) of the entire plant can usually be attributed to the utility and capital costs of its HEN (Verheyen and Zhang, 2006). Thus, the traditional aim of HEN synthesis consists of finding a cost-optimal network structure under the constraints that the operating conditions are all fixed. In the previous works on HEN flexibility, it is usually assumed that enough control loops have already been put in place to keep these process conditions at the nominal levels.

However, because there are always significant changes (uncertainties) in the plant environment (Verheyen and Zhang, 2006), dealing with uncertainties is an inherent feature of any HEN design. Marselle et al. (1982) pioneered the studies on HEN operability. It was proposed to manually integrate a series of optimal designs for different worst-case scenarios. Kotjabasakis and Linnhoff (1986) introduced the sensitivity tables for designing flexible HENs. To evaluate the operational flexibility of a HEN, Saboo et al. (1985) proposed to calculate the resilience index (RI), and Swaney and Grossmann (1985a, b) formulated the steady-state flexibility index mentioned in Chapter 2. Grossmann and Floudas (1987) then developed an active set strategy for the calculation of this index (also see Chapter 2). Subsequently, the multiperiod/multiscenario formulations were proposed by Grossmann and Floudas (1987), and the flexible HENs can be designed with a sequential approach. Papalexandri and Pistikopoulos (1994a, b) later formulated a large and complex MINLP model for the synthesis and retrofit of flexible and structurally controllable HENs.

7.2.2 MODEL FORMULATIONS

Because the topology of an HEN varies significantly from plant to plant, a generalized model formulation can be difficult to comprehend. Thus, a simple example is used here instead for illustration clarity. Specifically, let us consider the HEN presented in Figure 7.5 (Biegler et al., 1997). Based on the fact that only cooling is required in this network, one can deduce that its heat recovery level is maximized. Although such a design is economically attractive, its operational flexibility may not be acceptable. Let us assume that upstream disturbances may enter the inlet temperatures of cold stream T_3 and hot stream T_5. Let us also assume that their nominal values can be estimated to be 388 K and 583 K, respectively, and each may deviate ±10 K from these values.

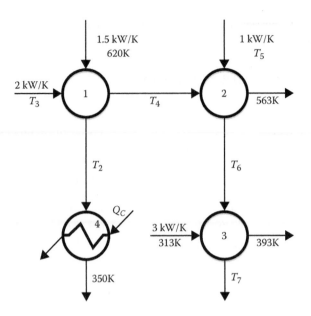

FIGURE 7.5   A minimum utility HEN design.

The equality and inequality constraints of the corresponding mathematical model can be respectively summarized as follows.

First of all, a heat balance equation can be established for each unit in Figure 7.5, that is,

$$\text{Exchanger 1: } 1.5(620 - T_2) = 2(T_4 - T_3) \tag{7.19}$$

$$\text{Exchanger 2: } (T_5 - T_6) = 2(563 - T_4) \tag{7.20}$$

$$\text{Exchanger 3: } (T_6 - T_7) = 3(393 - 313) \tag{7.21}$$

$$\text{Cooler}: Q_c = 1.5(T_4 - 350) \tag{7.22}$$

Second, the temperature differences at the hot and cold ends of every heat exchanger must be nonnegative, that is,

$$\text{Exchanger 1}: T_2 - T_3 \geq 0 \tag{7.23}$$

$$620 - T_4 \geq 0 \tag{7.24}$$

$$\text{Exchanger 2}: T_6 - T_4 \geq 0 \tag{7.25}$$

$$T_5 - 563 \geq 0 \tag{7.26}$$

$$\text{Exchanger 3}: T_7 - 313 \geq 0 \tag{7.27}$$

$$T_6 - 393 \geq 0 \tag{7.28}$$

Finally, the intermediate temperatures of each process stream must also be constrained as follows:

$$T_3 \leq T_4 \leq 563 \tag{7.29}$$

$$620 \geq T_2 \geq 350 \tag{7.30}$$

$$T_5 \geq T_6 \geq 323 \geq T_7 \tag{7.31}$$

Inequalities in Equations 7.23 through 7.28 essentially ensure that all heat exchanges are feasible. The intermediate temperatures between heat exchangers—T_4, T_2, and T_6—are bounded according to inequalities in Equations 7.29 through 7.31, and an additional constraint is imposed with Equation 7.31 on the target temperature of the second hot stream; that is, T_7 is allowed to reach any temperature that is lower than or equal to 323 K. The temperatures T_2, T_4, T_6, and T_7 can be regarded as the state variables with T_3 and T_5, being the uncertain parameters, and the cooler load Q_c is a control variable.

7.2.3 CASE STUDIES

If the heat load of the cooler (Q_c) remains unchanged, the corresponding steady-state flexibility index (FI_s) of the previously mentioned HEN can be easily calculated. Specifically, if Q_c is fixed at 75 kW, which is the cooler load at the nominal conditions of the uncertain parameters (i.e., $T_3^N = 388\ K$ and $T_5^N = 588\ K$), then $FI_s = 0.001$. Note that because these nominal temperatures lie almost at the boundary, as shown in Figure 7.6, FI_s is inevitably very small. On the other hand, the volumetric flexibility index could also be calculated as follows: $FI_v = 95.04 \div ((593 - 573) \times (398 - 378)) = 0.2376$. It can be clearly observed from Figure 7.6 that the feasible region of the HEN does not cover all points in the rectangle area formed by the upper and lower bounds of the uncertain parameters; therefore, $FI_v < 1$ in this case. However, this volumetric flexibility index still represents a much more optimistic assessment than that indicated by the steady-state flexibility index.

One may then want to see what happens if the control variable is adjusted to a different value (e.g., 60 kWh). It can be found in Figure 7.7 that the feasible region obtained from volumetric flexibility index analysis is shifted to the left. As a result, the area of the feasible region becomes smaller and $FI_v = 49.41 / 400 = 0.124$. This result indicates that the given HEN should be operable in certain conditions that are bounded by the feasible region, although the corresponding operating conditions may not be ideal (i.e., far from the designated operating condition, which one may expect to be in the middle of the specified uncertain region). On the other hand, FI_s is indeterminable in this case because the nominal operating condition is way outside the feasible region. It can thus be observed that FI_s may be grossly misinterpreted if the selected nominal point is not within the feasible region (Figure 7.8).

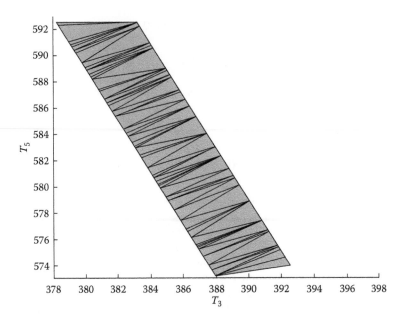

FIGURE 7.6   Feasible region of HEN with $Q_c = 75$ kW.

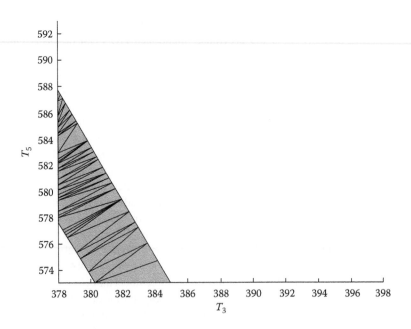

FIGURE 7.7   Feasible region of HEN with $Q_c = 60$ kW.

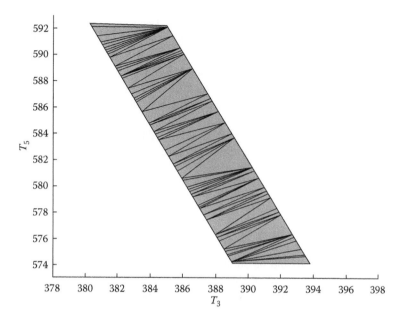

FIGURE 7.8   Feasible region of HEN with $Q_c = 80$ kW.

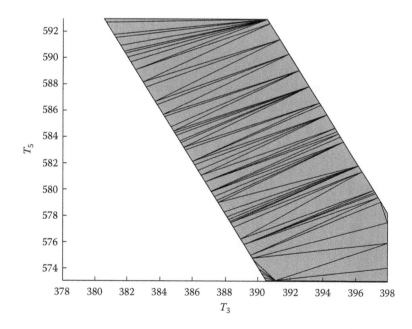

FIGURE 7.9   Relaxed feasible region of HEN with $Q_c = 80$ kW.

Finally, if Q_c is adjusted to 80 kW, the nominal condition is now somewhere in the middle of the feasible region, and thus, FI_s can be increased to 0.15. On the other hand, FI_v does not vary significantly (i.e., $FI_v = 94 / 400 = 0.235$). Notice that although FI_s is very sensitive to the chosen location of the nominal point, FI_v is only affected by the area of the feasible region. The only way to improve FI_v is by relaxing the temperature constraints so that the feasible region could be enlarged, for example, increasing the constraint temperature in Equation 7.31 from 323 K to 333 K. Although the corresponding FI_s is not altered because the active constraint is the same (see the left region boundary of Figure 7.9), FI_v can be increased significantly to $192.5 / 400 = 0.48$.

7.2.4 Concluding Remarks

The operational flexibility of an HEN could be analyzed using either steady-state or volumetric flexibility indices. In cases when the former yields an overly pessimistic assessment, the latter should be adopted as a complementary measure to improve design. The boundary points of the feasible region, which are accurately identified in volumetric flexibility analysis, could be adopted for pinpointing the most constrained segment of the active constraint(s) in steady-state flexibility analysis. Moreover, instead of the single set of nominal conditions adopted in the traditional ad hoc approach, many other options may be considered because any operable point within the aforementioned feasible region can be chosen for design.

REFERENCES

Aaltola, J., 2002. Simultaneous synthesis of flexible heat exchanger network. *Applied Thermal Engineering* 22, 907–918.

Abejon, R., Garea, A., Irabien, A., 2014. Analysis and optimization of continuous organic solvent nanofiltration by membrane cascade for pharmaceutical separation. *Aiche Journal* 60, 931–948.

Adi, V.S.K., Laxmidewi, R., Chang, C.T., 2016. An effective computation strategy for assessing operational flexibility of high-dimensional systems with complicated feasible regions. *Chemical Engineering Science* 147, 137–149.

Adler, S., Beaver, E., Bryan, P., Robinson, S., Watson, J., 2000. *Vision 2020:2000 Separations Roadmap.* New York, NY: Center for Waste Reduction Technologies of the AIChE.

Biegler, L.T., Grossmann, I.E., Westerberg, A.W., 1997. *Systematic Methods of Chemical Process Design.* Upper Saddle River, NJ: Prentice Hall PTR.

Geens, J., De Witte, B., Van der Bruggen, B., 2007. Removal of API's (Active Pharmaceutical Ingredients) from organic solvents by nanofiltration. *Separation Science and Technology* 42, 2435–2449.

Grossmann, I.E., Floudas, C.A., 1987. Active constraint strategy for flexibility analysis in chemical processes. *Computers & Chemical Engineering* 11, 675–693.

Kim, J.F., Szekely, G., Valtcheva, I.B., Livingston, A.G., 2014. Increasing the sustainability of membrane processes through cascade approach and solvent recovery-pharmaceutical purification case study. *Green Chemistry* 16, 133–145.

Kotjabasakis, E., Linnhoff, B., 1986. Sensitivity tables for the design of flexible processes.1. How much contingency in heat-exchanger networks is cost-effective. *Chemical Engineering Research & Design* 64, 197–211.

Lai, S.M., Hui, C.W., 2008. Process flexibility for multivariable systems. *Industrial & Engineering Chemistry Research* 47, 4170–4183.

Lin, J.C.T., Livingston, A.G., 2007. Nanofiltration membrane cascade for continuous solvent exchange. *Chemical Engineering Science* 62, 2728–2736.

Marchetti, P., Livingston, A.G., 2015. Predictive membrane transport models for organic solvent nanofiltration: How complex do we need to be? *Journal of Membrane Science* 476, 530–553.

Marselle, D.F., Morari, M., Rudd, D.F., 1982. Design of resilient processing plants.2. Design and control of energy management-systems. *Chemical Engineering Science* 37, 259–270.

Papalexandri, K.P., Pistikopoulos, E.N., 1994a. Synthesis and retrofit design of operable heat-exchanger networks.1. Flexibility and structural controllability aspects. *Industrial & Engineering Chemistry Research* 33, 1718–1737.

Papalexandri, K.P., Pistikopoulos, E.N., 1994b. Synthesis and retrofit design of operable heat-exchanger networks.2. Dynamics and control-structure considerations. *Industrial & Engineering Chemistry Research* 33, 1738–1755.

Saboo, A.K., Morari, M., Woodcock, D.C., 1985. Design of resilient processing plants.8. A resilience index for heat-exchanger networks. *Chemical Engineering Science* 40, 1553–1565.

Siew, W.E., Livingston, A.G., Ates, C., Merschaert, A., 2013a. Continuous solute fractionation with membrane cascades—A high productivity alternative to diafiltration. *Separation and Purification Technology* 102, 1–14.

Siew, W.E., Livingston, A.G., Ates, C., Merschaert, A., 2013b. Molecular separation with an organic solvent nanofiltration cascade—Augmenting membrane selectivity with process engineering. *Chemical Engineering Science* 90, 299–310.

Swaney, R.E., Grossmann, I.E., 1985a. An index for operational flexibility in chemical process design.1. Formulation and theory. *Aiche Journal* 31, 621–630.

Swaney, R.E., Grossmann, I.E., 1985b. An index for operational flexibility in chemical process design.2. Computational algorithms. *Aiche Journal* 31, 631–641.

Vandezande, P., Gevers, L.E.M., Vankelecom, I.F.J., 2008. Solvent resistant nanofiltration: Separating on a molecular level. *Chemical Society Reviews* 37, 365–405.

Verheyen, W., Zhang, N., 2006. Design of flexible heat exchanger network for multi-period operation. *Chemical Engineering Science* 61, 7730–7753.

Wijmans, J.G., Baker, R.W., 1995. The solution-diffusion model—A review. *Journal of Membrane Science* 107, 1–21.

... ...

8 Flexible Designs of Solar-Driven Membrane Distillation Desalination Systems

8.1 BACKGROUND

Due to the alarming effects of global warming and a growing world population, there is an ever-increasing demand on water resources almost everywhere on earth. Consequently, considerable research effort has been devoted to the development of an efficient and sustainable desalination technology in recent years. Among all viable alternatives, the air gap membrane distillation (AGMD) is widely considered a promising candidate because the energy consumed per unit of water generated is the lowest (Ben Bacha et al., 2007; Bui et al., 2010; Cabassud and Wirth, 2003). Many researchers have already rigorously analyzed the underlying transport phenomena to identify the key variables affecting the water flux in an AGMD module (Ben Bacha et al., 2007; Chang et al., 2010; Koschikowski et al., 2003; Meindersma et al., 2005, 2006). In particular, Ben Bacha et al. (2007) and Chang et al. (2010, 2012) have built models of all units embedded in a solar-driven membrane distillation desalination system (SMDDS): (1) the solar absorber, (2) the thermal storage tank, (3) the counterflow shell-and-tube heat exchanger, (4) the AGMD modules, and (5) the distillate tank and discussed various operational and control issues accordingly. A typical process flow diagram of an SMDDS can be found in Figure 8.1. Galvez et al. (2009) meanwhile designed a 50 cubic-meters-per-day desalination setup with an innovative solar-powered membrane, and Guillen-Burrieza et al. (2011) also assembled a solar-driven AGMD pilot. These two studies were performed with the common goal of minimizing the energy needed for producing one unit of distillate. It should be noted that although applications of the solar-driven AGMD modules were successful, the aforementioned works focused only upon thermal efficiency and the important issues concerning operational flexibility have not been addressed.

8.2 UNIT MODELS

The SMDDS units—that is, the solar absorber, the thermal storage tank, the counterflow shell-and-tube heat exchanger, the AGMD modules, and the distillate tank—are interconnected to form two distinct processing routes for seawater desalination and solar energy conversion, respectively. Obviously a realistic system design must be fully functional in the presence of uncertain sunlight radiation and unpredictable freshwater demand. To achieve a desired flexibility target, the aforementioned units must be sized properly and also the corresponding thermal storage scheme must be synthesized in a rational fashion. If the solar absorber is relatively small when compared with the membrane distillation unit, then it may be beneficial to operate the stripped-down SMDDS shown in Figure 8.1 (Structure I). Otherwise, at least one thermal storage tank must be adopted to buffer the drastic energy surplus incurred during daytime operation. Structure II in Figure 8.2 is the simplest design for such a purpose.

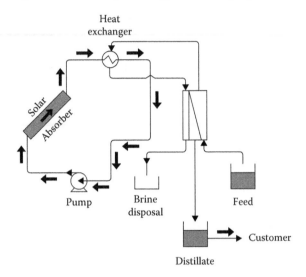

FIGURE 8.1   Structure I. (Reprinted from *Desalination*, Vol. 320, Adi, V.S.K., Chang, C.T., SMDDS design based on temporal flexibility analysis, 96–104. Copyright 2013 with permission from Elsevier.)

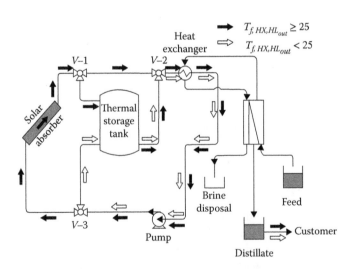

FIGURE 8.2   Structure II. (Reprinted from *Desalination*, Vol. 320, Adi, V.S.K., Chang, C.T., SMDDS design based on temporal flexibility analysis, 96–104. Copyright 2013 with permission from Elsevier.)

For implementation convenience, the available unit models (Chang et al., 2010) have been simplified as follows.

8.2.1 SOLAR ABSORBER

The solar absorber in an SMDDS design is used to convert solar energy to heat. The following assumptions are adopted in formulating its model: (1) the fluid velocities in all absorber tubes are the same; (2) the fluid temperature should be kept below the boiling point; (3) there is no water loss; and (4) heat loss is negligible. The corresponding transient energy balance can be written as

$$\frac{dT_{f,SA_{out}}}{dt} = -\frac{m_{f,SA}}{M_{f,SA}}\left(T_{f,SA_{out}} - T_{f,SA_{in}}\right) + \frac{A_{SA}I(t)}{M_{f,SA}Cp_f^L} \tag{8.1}$$

$$T_{f,SA_{out}} \le T_{f,SA_{out}}^{max} \tag{8.2}$$

where $T_{f,SA_{in}}$ and $T_{f,SA_{out}}$ denote the inlet and outlet temperatures (°C) of the solar absorber, respectively; $T_{f,SA_{out}}^{max}$ is the maximum allowable outlet temperature (°C); $M_{f,SA}$ denotes the total mass of operating fluid in the solar absorber (kg); $m_{f,SA}$ denotes the overall mass flow rate of operating fluid in solar absorber (kg/h); A_{SA} is the exposed area of solar absorber (m^2); Cp_f^L is the heat capacity of the operating fluid (J/kg°C); and $I(t)$ is the solar irradiation rate per unit area (W/m^2) at time t.

8.2.2 THERMAL STORAGE TANK

Notice that the thermal storage tank is present only in structure II (see Figure 8.2). By assuming that (1) the fluid inside the thermal storage tank is well mixed, (2) the inlet and outlet flow rates are identical, and (3) the heat capacity of the operating fluid is independent of temperature, the corresponding transient energy balance can be expressed as

$$M_{f,ST}\frac{dT_{f,ST_{out}}}{dt} = r_{f,ST}m_{f,STL}\left(T_{f,ST_{in}} - T_{f,ST_{out}}\right) \tag{8.3}$$

$$r_{f,ST} = \frac{m_{f,ST}}{m_{f,STL}} \tag{8.4}$$

where $T_{f,ST_{in}}$ and $T_{f,ST_{out}}$ denote the inlet and outlet temperatures (°C) of the thermal storage tank, respectively; $M_{f,ST}$ represents the total mass of operating fluid in the thermal storage tank (kg); $m_{f,STL}$ is the total mass flow rate driven by the pump in the thermal loop (kg/h); and $m_{f,ST}$ is the throughput of the thermal storage tank (kg/h), which equals $r_{f,ST}m_{f,STL}$.

For simplicity, let us assume that the solar absorber is disconnected from the thermal loop only when the outlet temperature of the hot fluid from the heat exchanger is lower than 25°C. In other words,

$$m_{f,SA} = \begin{cases} m_{f,STL} & \text{if } T_{f,HX,HL_{out}} \geq 25 \\ 0 & \text{if } T_{f,HX,HL_{out}} < 25 \end{cases} \qquad (8.5)$$

Also, the flow ratio $r_{f,ST}(t)$ defined in Equation 8.4 is treated as an adjustable quantity in structure II, that is, $0 \leq r_{f,ST}(t) \leq 1$, while held unchanged, respectively, at different levels in n finite time intervals; that is,

$$\frac{dr_{f,ST}}{dt} = 0 \qquad (8.6)$$

where $t_{i-1} < t < t_i$ ($i = 1, 2, \cdots, n$) and $0 = t_0 < t_1 < \cdots < t_n = H$. This practice can be justified on the basis of the argument that in actual implementation, a piecewise-constant control profile is more realizable than the time-variant counterpart implied by Karush-Kuhn-Tucker (KKT) condition (iv) derived in Section 4.2.3. Finally, in cases where the thermal storage tank is not utilized (i.e., structure I), one could simply set $m_{f,SA} = m_{f,STL}$ and $r_{f,ST} = 0$.

8.2.3   HEAT EXCHANGER

The hot fluid used in the counterflow heat exchanger comes from the thermal storage tank and/or solar absorber, whereas the cold fluid is the seawater. By assuming no heat loss and ignoring the transient behavior, a steady-state energy balance is used to characterize the heat exchange approximately. Thus, its unit model can be written as

$$m_{f,MD}\left(T_{f,HX,CL_{out}} - T_{f,HX,CL_{in}}\right) = m_{f,HX,HL}\left(T_{f,HX,HL_{in}} - T_{f,HX,HL_{out}}\right) \qquad (8.7)$$

where $m_{f,HX,HL}$ is the mass flow rate of hot fluid (kg / h); $T_{f,HX,HL_{in}}$ and $T_{f,HX,HL_{out}}$, respectively, denote the inlet and outlet temperatures of the hot fluid (°C); $m_{f,MD}$ is the mass flow rate of seawater in a membrane distillation loop (kg / h); and $T_{f,HX,CL_{in}}$ and $T_{f,HX,CL_{out}}$, respectively, denote the inlet and outlet temperatures of the cold fluid (°C). Note that the mass flow rate of hot fluid is essentially the same as that in the thermal loop in either structure I or structure II, that is,

$$m_{f,HX,HL} = m_{f,STL} \qquad (8.8)$$

An energy balance around the valve V-2 yields

$$T_{f,HX,HL_{in}} = \left(1 - r_{f,ST}\right)T_{f,SA_{out}} + r_{f,ST}T_{f,ST_{out}} \qquad (8.9)$$

Again, this equation is also valid in structure I when $r_{f,ST} = 0$. Finally, let us consider the outlet temperature of hot fluid. Because in structure II, the hot fluid leaving the heat exchanger is recycled either back to the solar absorber or directly to the thermal storage tank, the following constraints should be imposed:

$$T_{f,HX,HL_{out}} = \begin{cases} T_{f,SA_{in}} & \text{if } T_{f,HX,HL_{out}} \geq 25 \\ T_{f,ST_{in}} & \text{if } T_{f,HX,HL_{out}} < 25 \end{cases} \tag{8.10}$$

On the other hand, because structure I is not equipped with a thermal storage tank, only the first constraint in Equation 8.10 can be used in the corresponding model.

8.2.4 AGMD MODULE

To relieve the computation load, only a simplified model is adopted in this study for characterizing the AGMD unit. It is assumed that the mass flux of distillate across the membrane is a function of the energy input rate. Specifically, this flux in a standard module can be expressed as

$$N_{mem} = \frac{m_{f,MD} Cp_f^L \left(T_{f,HX,CL_{out}} - T_{f,HX,CL_{in}} \right)}{STEC \cdot A_{MD} \cdot n_{AGMD}} \tag{8.11}$$

where N_{mem} denotes the distillate flux (kg / m²h); A_{MD} is the fixed membrane area of a standard AGMD module (m²); n_{AGMD} is the total number of standard modules; and STEC is the *specific thermal energy consumption* constant (kJ / kg), which can be considered the ratio of the energy supplied by the heat exchanger to the mass of the distillate produced (Banat et al., 2007; Burgess and Lovegrove, 2016).

Strictly speaking, the mass flux through the AGMD membrane should be driven primarily by the vapor pressure differential. However, this flux is assumed here to be roughly proportional to the temperature difference for the purpose of simplifying the calculation. Because Equation 8.11 is used essentially as an empirical relation, in this case, it should be only valid within a finite range of the seawater flow rate. Consequently, $m_{f,MD}$ is treated in this work as a control variable that is allowed to vary ±10% from its nominal value, that is,

$$0.9 m_{f,MD}^N \leq m_{f,MD} \leq 1.1 m_{f,MD}^N \tag{8.12}$$

For the purpose of generating a realizable profile, this control variable is again kept unchanged at different levels in n distinct time intervals, that is,

$$\frac{dm_{f,MD}}{dt} = 0 \tag{8.13}$$

where $t_{i-1} < t < t_i$ $(i = 1, 2, \cdots, n)$ and $0 = t_0 < t_1 < \cdots < t_n = H$. Finally, note that the temperature of seawater entering the AGMD module should not be allowed to exceed a specified upper bound so as to avoid damaging the membrane, that is,

$$T_{f,HX,CL_{out}} \leq T_{f,HX,CL_{out}}^{max} \tag{8.14}$$

where $T_{f,HX,CL_{out}}^{max}$ is the upper bound of the cold stream temperature at the outlet of the heat exchanger (°C).

8.2.5 DISTILLATE TANK

The distillate tank is used as a buffer to ensure uninterrupted supply to the fluctuating water demand. The corresponding model can be written as

$$\rho_f^L A_{DT} \frac{dh_{DT}}{dt} = m_{f,DT_{in}} - m_{f,DT_{out}} \tag{8.15}$$

where ρ_f^L is the distillate density (kg / m³); A_{DT} is the cross-sectional area of the distillate tank (m²); h_{DT} is the height of liquid in the distillate tank (m); and $m_{f,DT_{in}}$ and $m_{f,DT_{out}}$ denote the inlet and outlet flow rates, respectively (kg / h). Note that the inlet flow is produced by the AGMD unit, that is,

$$m_{f,DT_{in}} = n_{AGMD} N_{mem} A_{MD} \tag{8.16}$$

Finally, it is evident that the liquid height in the distillate tank should be maintained within a specified range, that is,

$$h_{DT,low} \leq h_{DT} \leq h_{DT,high} \tag{8.17}$$

where $h_{DT,low}$ and $h_{DT,high}$, respectively, denote the given lower and upper bounds (m).

8.3 MODIFIED KKT CONDITIONS CONCERNING CONTROL VARIABLES

Due to the additional constraints (i.e., Equations 8.6 and 8.13) imposed upon the control variables, the corresponding KKT conditions for the present application should be modified slightly. Let us revisit the derivation presented in Section 4.2.3. The original necessary conditions in (i) to (iii) can still be produced by taking the first variation of the aggregated objective functional L defined in Equation 4.25 and then setting it to zero, whereas the last set of conditions can be obtained by considering the remaining term in δL after imposing the first three:

$$\delta L = \int_0^H \delta z \left[\mu^T \left(\frac{\partial \varphi}{\partial z} \right) + \lambda^T \left(\frac{\partial g}{\partial z} \right) \right] dt = \sum_{i=1}^n \delta z_i \int_{t_{i-1}}^{t_i} \left[\mu^T \left(\frac{\partial \varphi}{\partial z_i} \right) + \lambda^T \left(\frac{\partial g}{\partial z_i} \right) \right] dt = 0 \tag{8.18}$$

Because the elements in vector z_i (and also δz_i) are constants in $[t_{i-1}, t_i]$ but can be chosen independently and arbitrarily, Equation 8.18 implies that

$$\int_{t_{i-1}}^{t_i} \left[\mu^T \left(\frac{\partial \varphi}{\partial z_i} \right) + \lambda^T \left(\frac{\partial g}{\partial z_i} \right) \right] dt = 0 \tag{8.19}$$

where $i = 1, 2, \cdots, n$. Therefore, the KKT conditions in the present application should be those in sets (i), (ii), (iii)', and Equation 8.19.

8.4 CASE STUDIES

The case studies presented here are used mainly to show the usefulness of flexibility indices in SMDDS design. In all examples considered in this section, the specifications of a standard AGMD module are the same as those given in Banat et al. (2007). The effective area of the membrane is 10 m^2 per module. The flow channels in this module are fabricated by spiral-winding the membrane and condenser foils. The effluent of cold seawater flows into the shell side of a heat exchanger then into the hot flow channel of the AGMD unit. Because of the hydrophobic nature of the porous membrane, only water vapor passes through the membrane pore, this flux is driven primarily by the partial pressure difference across the membrane (Banat et al., 2007). The transported water vapor is condensed on the wall surface of the cold seawater flow channel and then collected in a distillate tank for consumption.

From Figures 8.1 and 8.2, it is quite obvious that the AGMD desalination unit is driven by the thermal energy carried in the operating fluid. In the daytime operation, the heat generated by the solar absorber can be consumed entirely in either structure I or II if the irradiation level is low. In the case of strong sunlight, a portion of the absorbed energy can be stored in the thermal storage tank of the second configuration then used later to facilitate desalination operation after sunset. Because the first structure is not equipped with an energy storage facility, it is necessary to use a relatively small absorber so as to ensure complete consumption of solar energy in the daytime and satisfy the freshwater demand during the night with the inventory stored in a properly sized distillate tank.

The solar irradiation rate $I(t)$ is regarded as a time-variant uncertain parameter in the flexibility analysis. Its nominal profile $I^N(t)$ and the expected upper and lower bounds are all depicted in Figure 8.3. Note that the expected positive and negative deviations at any time are both set at 10% of the nominal level. The water demand

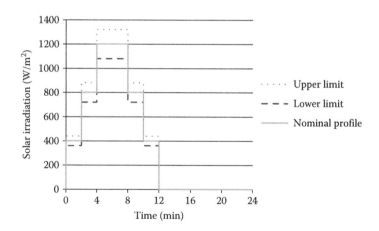

FIGURE 8.3 Solar irradiation rate. (Reprinted from *Desalination*, Vol. 320, Adi, V.S.K., Chang, C.T., SMDDS design based on temporal flexibility analysis, 96–104. Copyright 2013 with permission from Elsevier.)

rate $m_{f,DT_{out}}(t)$ is another time-dependent uncertain parameter considered in the case studies. Its nominal value is set at $18 \times \text{wdf}(t)$ kg / h, where $\text{wdf}(t)$ is the ratio between the demand rate at time t and the constant reference value of 18 kg / h. The expected deviations in $m_{f,DT_{out}}$ are also selected to be 10% of its nominal value. The nominal level of $\text{wdf}(t)$ and the corresponding upper and lower limits are sketched in Figure 8.4. It is assumed that the transient household water consumption rate can be closely characterized by the nominal profile of $\text{wdf}(t)$. Finally, it should be noted that if alternative solar irradiation profiles and water demand profiles can be made available in other applications, they can be easily incorporated in the proposed flexibility analysis so as to ensure realistic designs.

Before solving the proposed mathematical programs, all model parameters must be properly selected. Based on Equations 8.11 and 8.16, the production rate of each AGMD module at $T_{f,HX,CL_{out}} = 74°C$ is estimated to be 16.54 kg / h (Banat et al., 2007) (assuming that the feed temperature is $T_{f,HX,CL_{in}} = 25°C$). According to Banat et al. (2007), the nominal mass flow rate of seawater in a membrane distillation loop ($m_{f,MD}^{N}$) is 1,125 kg / h per AGMD module. Also, a maximum daily water consumption rate of 750.42 kg / day can be determined according to Figure 8.4. By adopting an average online period of 12 hour/day, the approximate number of parallel AGMD modules can be calculated, that is, $n_{AGMD} = \dfrac{750.42}{16.54 \times 12} = 3.78 \approx 4$, and thus the total membrane area should be 40 m². In the solar absorber, the total mass of operating fluid per unit area, that is, $M_{f,SA} / A_{SA}$, is set to be 15 kg / m² (Chang et al., 2010). The flow rate in the solar thermal loop ($m_{f,STL}$) is chosen to be 36,000 kg / h, which is eight times the total nominal flow rate of seawater in the membrane distillation loop ($m_{f,MD}^{N} = 1,125 \times 4 = 4,500$ kg / h). This value is selected to ensure a quick temperature response in the desalination loop. The volume of the distillate tank in each configuration is assumed to be 0.75 m³ ($A_{DT} = 0.35$ m²; $h_{DT,low} = 0$ m;

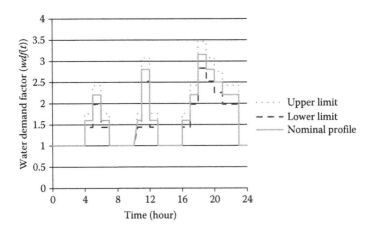

FIGURE 8.4 Water demand. (Reprinted from *Desalination*, Vol. 320, Adi, V.S.K., Chang, C.T., SMDDS design based on temporal flexibility analysis, 96–104. Copyright 2013 with permission from Elsevier.)

$h_{DT,high} = 2.14$ m), whereas a 10 m³ thermal storage tank ($M_{f,ST} = 10,000$ kg) is utilized in configuration II. Finally, it is assumed that the heat capacity of operating fluid Cp_f^L is held constant at 4,200 J / kg°C, and its density ρ_f^L is also assumed to be constant at 1,000 kg / m³.

As previously mentioned, the solar absorber should be sized according to the AGMD capacity. To facilitate a proper decision, an asymptotic energy utilization ratio between these two units can be defined for use as a rough measure of their size ratio:

$$\phi_{util} = \frac{\text{maximum supply rate of solar energy}}{\text{maximum consumption rate of thermal energy}}$$

$$= \frac{A_{SA}I^{max}}{m_{f,MD}^{max}Cp_f^L\left(T_{f,HX,CL_{out}}^{max} - T_{f,HX,CL_{in}}^{min}\right)} \tag{8.20}$$

Obviously, the energy captured by the solar absorber can be fully utilized by the AGMD module if this ratio is not larger than 1 ($\phi_{util} \leq 1$). From Figure 8.3, it can be observed that $I^{max} = 1,320$ W / m². On the basis of Equation 8.12, one could deduce $m_{f,MD}^{max} = 1.1 \times m_{f,MD}^{N} = 1,237.5$ kg / h. Also, from the previous model description, it is reasonable to assume that $T_{f,HX,CL_{out}}^{max} = 100$ °C and $T_{f,HX,CL_{in}}^{min} = 25$ °C. Note that only a simple calculation is needed to size the solar absorber according to a given ϕ_{util}. For example, the absorber area for $\phi_{util} = 1$ should be

$$A_{SA} = \frac{1,237.5 \times 4 \times 4,200 \times (100 - 25)}{1,320 \times 3,600} = 328.13 \text{ m}^2.$$ For the sake of completeness, all model parameters and variables used in the case studies are also listed in Table 8.1. For ease of implementation, the two control variables in the preliminary case studies

TABLE 8.1
Nomenclature

Symbol	Definition	Value	Classification
$T_{f,SA_{out}}^{max}$	Maximum allowable outlet temperature of the solar absorber	100 °C	d
$M_{f,SA}$	Total mass of operating fluid in the solar absorber	–	d
A_{SA}	Exposed area in the solar absorber	–	d
Cp_f^L	Heat capacity of the operating fluid	4,200 J / kg°C	d
$M_{f,ST}$	Total mass of operating fluid in the thermal storage tank	10,000 kg	d
$m_{f,STL}$	Mass flow rate in the thermal loop	36,000 kg / h	d
$T_{f,HX,CL_{in}}$	Cold fluid inlet temperature of the heat exchanger	25 °C	d
A_{MD}	Membrane area of a standard AGMD module	10 m²	d
n_{AGMD}	Total number of standard AGMD modules	4	d
STEC	Specific thermal energy consumption	14,000 kJ / kg	d

(Continued)

TABLE 8.1 (*Continued*)
Nomenclature

Symbol	Definition	Value	Classification
$T_{f,HX,CL_{out}}^{max}$	Maximum cold fluid outlet temperature of the heat exchanger	100 °C	d
ρ_f^L	Distillate density	1,000 kg / m^3	d
A_{DT}	Cross-sectional area of the distillate tank	0.35 m^2	d
$h_{DT,low}$	Lower bound of liquid height in the distillate tank	0 m	d
$h_{DT,high}$	Upper bound of liquid height in the distillate tank	2.14 m	d
ϕ_{util}	Energy utilization ratio	To be selected	d
I^{max}	Maximum solar irradiation rate per unit area	1,320 W / m^2	d
$m_{f,MD}^{max}$	Maximum mass flow rate in the membrane distillation loop	1,237.5 kg / h	d
$T_{f,HX,CL_{in}}^{min}$	Minimum cold fluid inlet temperature of the heat exchanger	25 °C	d
$T_{f,SA_{in}}$	Inlet temperature of the solar absorber	–	
$T_{f,SA_{out}}$	Outlet temperature of the solar absorber	–	x
$m_{f,SA}$	Mass flow rate of operating fluid in the solar absorber	–	x
$T_{f,ST_{in}}$	Inlet temperature of the thermal storage tank	–	x
$T_{f,ST_{out}}$	Outlet temperature of the thermal storage tank	–	x
$m_{f,ST}$	Throughput of the thermal storage tank	–	x
$m_{f,HX,HL}$	Mass flow rate of hot fluid in the heat exchanger	–	x
$T_{f,HX,HL_{in}}$	Hot fluid inlet temperature of the heat exchanger	–	x
$T_{f,HX,HL_{out}}$	Hot fluid outlet temperature of the heat exchanger	–	x
$T_{f,HX,CL_{out}}$	Cold fluid outlet temperature of the heat exchanger	–	x
h_{DT}	Liquid height in distillate tank	–	x
$m_{f,DT_{in}}$	Inlet flow rate of distillate tank	–	x
N_{mem}	Distillate flux through the AGMD membrane	–	x
$m_{f,MD}$	Mass flow rate in the membrane distillation loop	4,500 kg/h (nominal)	z
$r_{f,ST}$	Flow ratio for the thermal storage tank	–	z
$I(t)$	Solar irradiation rate per unit area	–	θ
$m_{f,DT_{out}}(t)$	Outlet flow rate of the distillate tank	–	θ

Source: Reprinted from *Desalination*, Vol. 320, Adi, V.S.K., Chang, C.T., SMDDS design based on temporal flexibility analysis, 96–104. Copyright 2013 with permission from Elsevier.

are both kept unchanged throughout the entire time horizon, that is, there is only one time interval ($n = 1$). A systematic approach is followed in these studies to size the solar absorber on the basis of Equation 8.20 and a given AGMD module size. By adopting the aforementioned thermal storage tank and distillate tank, the flexibility indices of structures I and II can be computed according to different utilization ratios (Adi and Chang, 2012). A summary of the optimization results is provided in Table 8.2 a through d.

It can be seen from Table 8.2 that when $\phi_{util} < 1$, both configurations yield the same flexibility indices. This is because of the fact that the absorbed solar energy is consumed almost immediately and completely; the thermal storage tank in structure II is not needed at all (i.e., $r_{f,ST}(t) = 0$). On the other hand, one can see that $r_{f,ST}(t) \geq 0$ if $\phi_{util} > 1$, which implies that the thermal storage tank is utilized for storing the excess solar energy acquired during daytime operation in structure II. Note also that the active constraint in each solution, that is, when $g_j = 0$, is also given in Table 8.2 a through d. In the cases when $\phi_{util} < 1$ is chosen, because the consumed energy may not be enough to meet the demand, the distillate tank is expected to be emptied at some instances. The optimization results of the corresponding two cases are analyzed as follows:

TABLE 8.2a
Optimization Results in Preliminary Case Studies

Structure	Case	1	2	3	4	5	6	7
	ϕ_{util}	0.683	0.75	1	1.04	1.112	1.25	1.34
I	FI_d	0	0.415	1.077	0.664	0	infeasible	infeasible
	$m_{f,MD}$	4050	4050	4950	4950	4950	N/A	N/A
	$g_j = 0$	$h_{DT,low}$	$h_{DT,low}$	$T_{f,SA_{out}}^{max}$	$T_{f,SA_{out}}^{max}$	$T_{f,SA_{out}}^{max}$	N/A	N/A

Source: Reprinted with permission from Kuo 2015. Copyright 2015 National Cheng Kung University Library.

TABLE 8.2b
Optimization Results in Preliminary Case Studies

Structure	Case	1	2	3	4	5	6	7
	ϕ_{util}	0.683	0.75	1	1.04	1.112	1.25	1.34
I	FI_t	0	0.723	†	0.242	0	infeasible	infeasible
	$m_{f,MD}$	4050	4050	†	4950	4950	N/A	N/A
	$g_j = 0$	$h_{DT,low}$	$h_{DT,low}$	†	$T_{f,SA_{out}}^{max}$	$T_{f,SA_{out}}^{max}$	N/A	N/A

Source: Reprinted with permission from Kuo 2015. Copyright 2015 National Cheng Kung University Library.

† Unnecessary.

TABLE 8.2c
Optimization Results in Preliminary Case Studies

	Case	1	2	3	4	5	6	7
Structure								
	ϕ_{util}	0.683	0.75	1	1.04	1.112	1.25	1.34
II	FI_d	0	0.415	1.698	1.872	1.457	0.554	0
	$r_{f,ST}$	0	0	0	0	0.064	0.083	0.137
	$m_{f,MD}$	4050	4050	4050	4050	4841.76	4950	4950
	$g_j = 0$	$h_{DT,low}$	$h_{DT,low}$	$h_{DT,low}$	$h_{DT,low}$	$T^{max}_{f,SA_{out}}$	$T^{max}_{f,SA_{out}}$	$T^{max}_{f,SA_{out}}$
						$h_{DT,high}$	$h_{DT,high}$	$h_{DT,high}$

Source: Reprinted with permission from Kuo 2015. Copyright 2015 National Cheng Kung University Library.

TABLE 8.2d
Optimization Results in Preliminary Case Studies

	Case	1	2	3	4	5	6	7
Structure								
	ϕ_{util}	0.683	0.75	1	1.04	1.112	1.25	1.34
II	FI_t	0	0.723	†	†	†	0.389	0
	$r_{f,ST}$	0	0	†	†	†	1	0.137
	$m_{f,MD}$	4050	4050	†	†	†	4950	4950
	$g_j = 0$	$h_{DT,low}$	$h_{DT,low}$	†	†	†	$T^{max}_{f,SA_{out}}$	$T^{max}_{f,SA_{out}}$
							$h_{DT,high}$	$h_{DT,high}$

Source: Reprinted with permission from Kuo 2015. Copyright 2015 National Cheng Kung University Library.
† Unnecessary.

- Let us first consider Case 1 when $\phi_{util} = 0.683$. Note that $FI_d = 0$ and $FI_t = 0$, that is, no deviations from the nominal parameters are allowed for both configurations throughout the entire operation horizon. This is due to the fact that the nominal absorption rate of solar energy is just enough to meet the nominal demand.
- The dynamic flexibility indices of both structures when $\phi_{util} = 0.75$ are identical, that is, $FI_d = 0.415$, and the corresponding temporal indices are also the same, that is, $FI_t = 0.723$, in this case. The simulation results of the worst-case scenarios are plotted in Figures 8.5 and 8.6. Note in the latter figure that the distillate tank in either structure is just emptied at the end of 24 hours. It is also found that the worst-case scenario considered

in computing the temporal flexibility index is concerned with a negative deviation of the solar irradiation rate from its nominal value to its lower bound between 241 and 441 minutes and a positive deviation of the water demand from its nominal value to its upper bound between 938 and 1354 minutes at the same time. Clearly, structure I should be chosen in this case because the equipment cost of the thermal storage facility can be cut completely.

By raising the energy utilization ratio to 1 (i.e., $\phi_{util} = 1$), structure II can be made more flexible ($FI_d = 1.698$) than structure I ($FI_d = 1.077$) in Case 3. Notice that the corresponding active constraints are not the same. The outlet temperature of the solar absorber reached its upper bound after 8 hours in structure I, whereas the water level in the distillate tank dropped to its lower bound at the end of the horizon in the structure II. Notice also that a larger-than-1 FI_d implies that the given process is operable throughout the entire horizon and thus guarantees $FI_t > 1$. As a result, the computation of the temporal flexibility index is unnecessary.

Next, let us consider additional cases in which the solar absorbers are larger than 1, that is, $\phi_{util} > 1$. Following are the corresponding descriptions and discussions:

- Structure I in Case 4 ($\phi_{util} = 1.04$) obviously cannot withstand at least some of the disturbances characterized by Figures 8.3 and 8.4 because $FI_d = 0.664$, but the dynamic flexibility index of structure II ($FI_d = 1.872$) indicates otherwise. Note that the active constraints in these two systems are not the same either. The active constraint is associated with the upper bound of the outlet temperature of the solar absorber after 8 hours in the former case, whereas in the latter, the lower bound of the water level in the distillate tank is at the end of the time horizon. Furthermore, it can also be found that the worst-case scenario in evaluating the temporal flexibility index for structure I ($FI_t = 0.242$) should be a positive deviation of the solar irradiation rate from its nominal value to its upper bound between 414 and 480 minutes. The simulated time profiles of two critical variables—$T_{f,SA_{Out}}$ and h_{DT}—in the worst-case scenarios can be found in Figures 8.7 through 8.10.
- In Case 5 ($\phi_{util} = 1.112$), the dynamic flexibility indices of structures I and II were found to be 0 and 1.46, respectively. The two active constraints in the latter case are now associated with the upper bound of the water level in the distillate tank after 16 hours and the upper bound of the outlet temperature of the solar absorber after 8 hours. This is obviously due to the fact that the solar energy is introduced at a rate that is much faster than the consumption rate of thermal energy and that the water production rate is also higher than the water demand. On the other hand, note that the dynamic flexibility index for structure I is zero. This drastic reduction in flexibility can also be attributed to the high intake rate of solar energy. Because there is no thermal storage tank, it is very difficult to keep the outlet temperature of the solar absorber ($T_{f,SA_{out}}$) below 100°C.

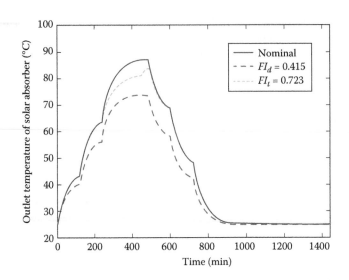

FIGURE 8.5   The time profiles of the solar absorber outlet temperature ($T_{f,SAout}$) for both structures in the worst-case scenario ($\phi_{util} = 0.75$). (Reprinted from *Journal of the Taiwan Institute of Chemical Engineers*, Wu, R.S., Chang., C.T., Development of mathematical programs for evaluating dynamic and temporal flexibility indices based on KKT conditions. Copyright 2017 with permission from Elsevier.)

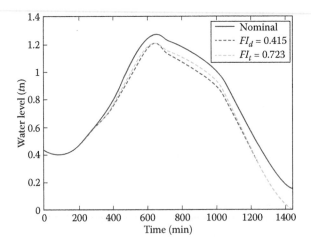

FIGURE 8.6   The time profiles of water levels in distillate tanks (h_{DT}) for both structures in the worst-case scenarios ($\phi_{util} = 0.75$). (Reprinted from *Journal of the Taiwan Institute of Chemical Engineers*, Wu, R.S., Chang, C.T., Development of mathematical programs for evaluating dynamic and temporal flexibility indices based on KKT conditions. Copyright 2017 with permission from Elsevier.)

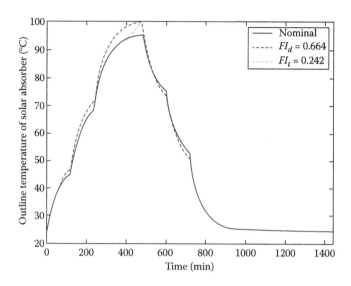

FIGURE 8.7 The time profiles of the solar absorber outlet temperature ($T_{f,SA_{out}}$) in the worst-case scenarios for structure I ($\phi_{util} = 1.04$). (Reprinted from *Journal of the Taiwan Institute of Chemical Engineers*, Wu, R.S., Chang, C.T., Development of mathematical programs for evaluating dynamic and temporal flexibility indices based on KKT conditions. Copyright 2017 with permission from Elsevier.)

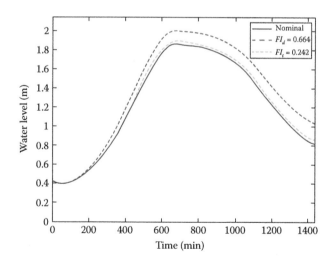

FIGURE 8.8 The time profiles of the water level in the distillate tank (h_{DT}) in the worst-case scenarios for structure I ($\phi_{util} = 1.04$). (Reprinted from *Journal of the Taiwan Institute of Chemical Engineers*, Wu, R.S., Chang, C.T., Development of mathematical programs for evaluating dynamic and temporal flexibility indices based on KKT conditions. Copyright 2017 with permission from Elsevier.)

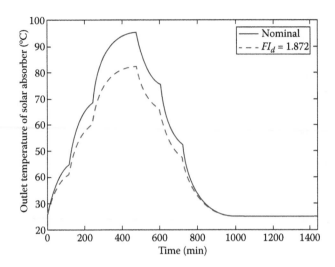

FIGURE 8.9   The time profiles of the solar absorber outlet temperature $(T_{f,SA_{out}})$ in the worst-case scenario of structure II $(\phi_{util} = 1.04)$. (Reprinted from *Journal of the Taiwan Institute of Chemical Engineers*, Wu, R.S., Chang, C.T., Development of mathematical programs for evaluating dynamic and temporal flexibility indices based on KKT conditions. Copyright 2017 with permission from Elsevier.)

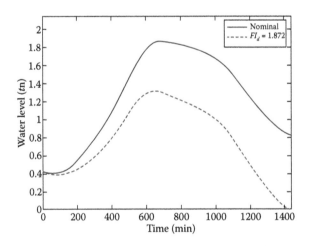

FIGURE 8.10   The time profile of the water level in the distillate tank (h_{DT}) in the worst-case scenario for structure II $(\phi_{util} = 1.04)$. (Reprinted from *Journal of the Taiwan Institute of Chemical Engineers*, Wu, R.S., Chang, C.T., Development of mathematical programs for evaluating dynamic and temporal flexibility indices based on KKT conditions. Copyright 2017 with permission from Elsevier.)

- In Case 6 ($\phi_{util} = 1.25$) and Case 7 ($\phi_{util} = 1.34$), the selected solar absorbers are larger than those used in the previous cases. Because more water is produced in structure II but the size of the distillate tank remains the same in either Case 6 or Case 7, the resulting flexibility index becomes much lower than that achieved in Case 5. Note that $FI_d = 0$ for structure I in Case 5 and for structure II in Case 7. Thus, any further increase in the utilization ratio in both cases inevitably renders the corresponding configuration infeasible.

By plotting the dynamic flexibility indices of structures I and II at various values of the asymptotic energy utilization ratio, one can construct Figure 8.11 and identify five regions (A to E) as shown. In regions A and B are designs indicating that the thermal storage tank is useless for enhancing flexibility. Thus, its budget can be eliminated completely. On the other hand, the thermal storage tank should be helpful in raising FI_d to a larger-than-1 value in regions C and D and making the resulting designs more flexible than structure I. Finally, designs in region E should be avoided because they are inflexible and the most expensive.

Thus, with sufficient funds, one can certainly choose the designs in regions B, C, and D to achieve the desired flexibility target. Otherwise, by relaxing the stringent criterion of $FI_d \geq 1$ the designs in region A can probably be selected after rigorous temporal flexibility analyses. Figure 8.12 shows the values of FI_t in region A, and clearly structure I with $\phi_{util} > 0.79$ should be a good candidate.

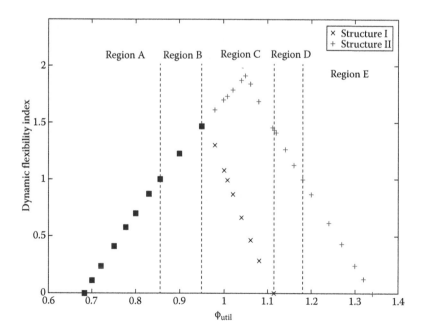

FIGURE 8.11 Dynamic flexibility indices of various system sizes.

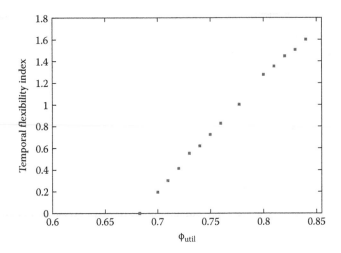

FIGURE 8.12   Temporal flexibility indices of structure I in Region A.

8.5   CONCLUDING REMARKS

As previously mentioned, the ability of a system to maintain feasible operation despite unexpected disturbances is referred to as its *operational flexibility*. A systematic SMDDS design strategy is thus developed in this work through flexibility analysis. Given a system configuration, all units can be appropriately sized to achieve a target degree of flexibility. Given a fixed SMDDS design, additional enhancement measures can be identified according to the active constraints embedded in the optimum solution of the flexibility index model. These measures for further refinements include modifications in unit sizes and/or system structure. Finally, the optimization and simulation results obtained in case studies show that the proposed approach is convenient and effective for addressing various operational issues in SMDDS design.

REFERENCES

Adi, V.S.K., Chang, C.T., 2012. Dynamic flexibility analysis with differential quadratures. *Computer Aided Chemical Engineering* 31, 260–264.

Adi, V.S.K., Chang, C.T., 2013. SMDDS design based on temporal flexibility analysis. *Desalination* 320, 96–104.

Banat, F., Jwaied, N., Rommel, M., Koschikowski, J., Wieghaus, M., 2007. Performance evaluation of the "large SMADES" autonomous desalination solar-driven membrane distillation plant in Aqaba, Jordan. *Desalination* 217, 17–28.

Ben Bacha, H., Dammak, T., Ben Abdalah, A.A., Maalej, A.Y., Ben Dhia, H., 2007. Desalination unit coupled with solar collectors and a storage tank: Modelling and simulation. *Desalination* 206, 341–352.

Bui, V.A., Vu, L.T.T., Nguyen, M.H., 2010. Simulation and optimisation of direct contact membrane distillation for energy efficiency. *Desalination* 259, 29–37.

Burgess, G., Lovegrove, K., 2016. Solar thermal powered desalination: Membrane versus distillation technologies. Centre for Sustainable Energy Systems, Department of Engineering. Australian National University, Canberra.

Cabassud, C., Wirth, D., 2003. Membrane distillation for water desalination: How to chose an appropriate membrane? *Desalination* 157, 307–314.

Chang, H., Lyu, S.G., Tsai, C.M., Chen, Y.H., Cheng, T.W., Chou, Y.H., 2012. Experimental and simulation study of a solar thermal driven membrane distillation desalination process. *Desalination* 286, 400–411.

Chang, H.A., Wang, G.B., Chen, Y.H., Li, C.C., Chang, C.L., 2010. Modeling and optimization of a solar driven membrane distillation desalination system. *Renewable Energy* 35, 2714–2722.

Galvez, J.B., Garcia-Rodriguez, L., Martin-Mateos, I., 2009. Seawater desalination by an innovative solar-powered membrane distillation system: The MEDESOL project. *Desalination* 246, 567–576.

Guillen-Burrieza, E., Blanco, J., Zaragoza, G., Alarcon, D.C., Palenzuela, P., Ibarra, M., Gernjak, W., 2011. Experimental analysis of an air gap membrane distillation solar desalination pilot system. *Journal of Membrane Science* 379, 386–396.

Koschikowski, J., Wieghaus, M., Rommel, M., 2003. Solar thermal-driven desalination plants based on membrane distillation. *Desalination* 156, 295–304.

Kuo, Y.C., 2015. Applications of the dynamic and temporal flexibility indices, Department of Chemical Engineering. National Cheng Kung University, Tainan.

Meindersma, G.W., Guijt, C.M., de Haan, A.B., 2005. Water recycling and desalination by air gap membrane distillation. *Environmental Progress* 24, 434–441.

Meindersma, G.W., Guijt, C.M., de Haan, A.B., 2006. Desalination and water recycling by air gap membrane distillation. *Desalination* 187, 291–301.

Wu, R.S., Chang, C.T., 2017. Development of mathematical programs for evaluating dynamic and temporal flexibility indices based on KKT conditions. *Journal of the Taiwan Institute of Chemical Engineers* 73, 86–92.

9 Flexible Designs of Hybrid Power Generation Systems for Standalone Applications

9.1 BACKGROUND

Renewable energy sources have been attracting strong interest in recent years and gained unprecedented importance after the COP 21 conference in Paris as viable alternatives to fossil fuels for power generation. Among the various technological options, those driven by sunlight, wind, and hydrogen appear to be mature enough for practical applications. The pros and cons of these different power-generating methods are briefly summarized in the sequel:

- The photovoltaic (PV) modules are capable of converting both direct and scattered sunlight into electricity. Although solar energy is inexhaustible and the power generation process is carbon free, it is widely recognized that the PV generator alone is not suitable for off-grid applications due to the intermittent and uncertain nature of sunlight irradiation. One way to overcome this problem is to complement it with at least one additional source (Sadri and Hooshmand, 2012).
- A wind turbine (WT) draws upon the force of moving air to generate electricity by rotating the propeller-like blades around a rotor. An electric network can certainly make use of the power produced by WT modules (Wang and Nehrir, 2008). Whether the demand is short term or long term, wind alone cannot provide power continuously due to its random speed and direction. However, if used in conjunction with other energy sources, it can offer economic benefits.
- The fuel cell (FC) is a quiet, responsive, and well-tested alternative for backup power (Hwang et al., 2008). However, the continuous runtime of an FC unit is often constrained by the onsite storage capacity of hydrogen. For sites with relatively low power loads and short outages lasting from hours to days, the fuel cell can be an ideal candidate.

It should also be noted that the operability of a PV or WT module can be greatly enhanced if it is augmented with batteries (Sadri and Hooshmand, 2012). However, it is only an acceptable choice for short-duration backup support due to the need for frequent replacement (Saathoff, 2016). The U.S. Department of Energy also suggested that a combination of more than one type of power-generating technology should, in principle, outperform any single-source system because their outputs peak at different times during the day and, therefore, the hybrid systems are more likely to produce power on demand (Schoenung, 2011). Based on these considerations, one can see that there is clearly a strong incentive to develop a systematic approach to producing practically feasible hybrid power solutions that incorporate all aforementioned power generation and energy storage options for off-grid applications.

Although numerous studies on the modeling, design, and optimization of photovoltaic–fuel cell–wind turbine (PVFCWT) systems have already been carried out in recent years—for example, see Banos et al. (2011), Li et al. (2009), and Zhou et al. (2010)—most of them ignored the important issues related to operational flexibility. Bajpai and Dash (2012) comprehensively reviewed studies concerning the hybrid systems for standalone applications, but none provided procedures that can accurately analyze their operational performance. Although incorporation of a large enough battery may smooth the power generation operation, it is still necessary to optimally allocate the capacities of various units in a hybrid system so as to avoid overdesign. For this purpose, the operational flexibility of any given system must be evaluated rigorously with one or more quantitative measures. It is worth noting that Erdinc and Uzunoglu (2012) have done a thorough review of the optimum design of hybrid renewable energy systems. A detailed analysis of different optimum sizing approaches was provided, but most, if not all, of the approaches do not take into account system flexibility.

Traditionally, the term *flexibility* is defined as the ability of a system to maintain feasible operation despite unexpected disturbances. Various approaches have been proposed to devise a metric to facilitate flexibility analysis (Adi and Chang, 2011; Bansal et al., 2000, 2002; Floudas et al., 2001; Grossmann and Floudas, 1987; Grossmann and Halemane, 1982; Halemane and Grossmann, 1983; Lima and Georgakis, 2008; Lima et al., 2010a, b; Malcolm et al., 2007; Ostrovski et al., 2002; Ostrovsky et al., 2000; Swaney and Grossmann, 1985a, b; Varvarezos et al., 1995; Volin and Ostrovskii, 2002). The original steady-state flexibility index (FI_s) was first defined by Swaney and Grossmann (1985a, b) for use as an unambiguous gauge of the feasible region in the uncertain parameter space. Specifically, the value of FI_s is associated with the maximum allowable deviations of the uncertain parameters from their nominal values, by which feasible operation can be assured with proper manipulation of the control variables. Because the steady-state material and energy balances are used as the equality constraints in this optimization problem (Grossmann and Floudas, 1987; Ostrovski et al., 2002; Ostrovsky et al., 2000; Swaney and Grossmann, 1985a, b; Varvarezos et al., 1995; Volin and Ostrovskii, 2002), the traditional steady-state flexibility index should be regarded as a performance measure of the continuous processes (Petracci et al., 1996; Pistikopoulos and Grossmann, 1988a, b, 1989a, b).

As indicated by Dimitriadis and Pistikopoulos (1995), the operational flexibility of a dynamic system should be evaluated differently. By adopting a system of differential algebraic equations (DAEs) as the model constraints, they developed a mathematical programming formulation for computing the dynamic flexibility index (FI_d). In addition, although the unexpected fluctuations in some process parameters may render an ill-designed system inoperable at certain instances, their cumulative effects can result in serious consequences as well. In a prior study, Adi and Chang (2013) developed a generic mathematical program to compute the temporal flexibility index (FI_t) for quantifying the system's ability to buffer the accumulated changes in uncertain parameters. Further work by Kuo and Chang (2016) reveals its important role in design. Specifically, FI_t could be used by the decision maker as a complementary criterion when FI_d is lower than the target value of 1. Because the sunlight irradiation, wind speed and direction, and hydrogen supply may all vary continuously with time and a PVFCWT system usually utilizes the battery units for storing the excess energy, both FI_d and FI_t may have to be computed to facilitate rigorous flexibility analyses.

In the previous work, Adi and Chang (2015) used only the temporal flexibility index to evaluate simple PVFC hybrid systems. More comprehensive studies are therefore needed to extend this approach further to design the PVFCWT hybrid systems by both dynamic and temporal flexibility indices. Also, because some well-developed software modules are readily available for simulating the realistic PV, FC, and battery units, a systematic optimization strategy has thus been developed in this work to integrate these existing codes for computing FI_d and FI_t. This simulation-based flexibility evaluation strategy is tested in a practical application for planning and designing the standalone hybrid power systems at the Annan campus of National Cheng Kung University in Tainan, Taiwan.

9.2 UNIT MODELS

As shown in Figure 9.1, four building blocks have been considered in the conventional analysis, that is, solar cell, wind turbine, polymer electrolyte membrane (PEM) fuel cell, and battery (Saathoff, 2016). The incorporation of the photovoltaic and/or wind modules is obviously designed to take advantage of the "free" energies as much as possible. During daylight hours, the load may be supported primarily by the former component, and as the wind blows, additional energy can be further extracted with the latter. For reliable power generation, these two modules should be complemented with batteries and perhaps also a fuel cell. If the power generated by the wind and solar energies exceeds the load, the excess current can be diverted toward the batteries and subsequently stored for later use. On the other hand, because the fuel cell is a more stable source, incorporation of such a unit minimizes the deep level of battery discharge and thus allows the hybrid system to function for a much longer period. Finally, it may also be necessary to include two more units in the proposed hybrid system to secure economical and reliable hydrogen supplies. A hydrogen storage facility is needed if only external sources are allowed, whereas an

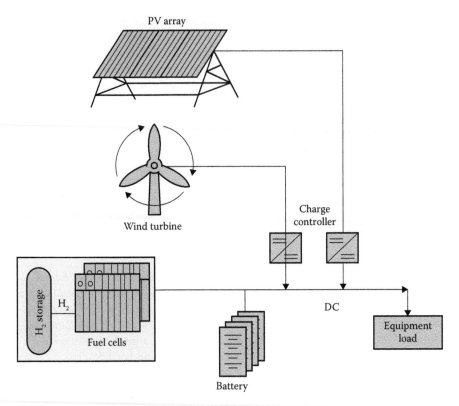

FIGURE 9.1 PV/FC/WT power generation system (From Saathoff, S., 2016. Hybrid power is the new green, OSP Magazine).

additional water electrolyzer should be considered to facilitate full recovery of the excess solar/wind energy.

The mathematical models and the corresponding simulation codes of those six units are briefly described next.

9.2.1 PHOTOVOLTAIC CELL

Let us roughly represent the essential characteristics at the terminals of practical arrays of interconnected PV cells as follows:

$$I_{PV} = I_{ph} - I_0 \left(\exp\left(\frac{V_{PV} + R_s I_{PV}}{V_t a} \right) - 1 \right) - \frac{V_{PV} + R_s I_{PV}}{R_p} \tag{9.1}$$

where I_{PV} and V_{PV} are the current (A) and voltage (V) of the PV module; I_{ph} and I_0 are the photovoltaic and saturation currents (A), respectively; $V_t = N_s kT/q$ is the thermal voltage (V) of the array with N_s cells connected in series; R_s is the equivalent series resistance (Ω); R_p is the equivalent parallel resistance (Ω); and a is the diode ideality constant. Because the detailed model formulation has already been presented elsewhere (Villalva et al., 2009), it is omitted here for the sake of conciseness. In fact, several well-developed computer programs and accurate, ready-to-use circuit models are already available for download at http://sites.google.com/site/mvillalva/pvmodel.

9.2.2 PEM Fuel Cell

The numerical simulation codes of the PEM fuel cell stacks according to the existing unit model (Wu et al., 2013) have also been developed. With an empirical formula (Khan and Iqbal, 2009; Wu et al., 2009), the following equation facilitates calculation of the output voltage of a single cell (V_{FC}):

$$V_{FC} = E - V_{act} - V_{ohm} \tag{9.2}$$

In an operating fuel cell, the actual voltage V_{FC} should be lower than its open-circuit voltage E due to various irreversible mechanisms that result in the activation overvoltage V_{act} and the ohmic overvoltage V_{ohm}.

Note that the Nernst equation results in an expression of the open circuit cell potential E, that is,

$$E = 1.229 - 8.5 \times 10^{-4}(T_{FC} - 298.15) + \frac{RT_{FC}}{2F} \ln\left(P_{H_2}P_{O_2}^{1/2}\right) \tag{9.3}$$

where R denotes the gas constant whose value is 8.314 J K^{-1}mol^{-1}; P_{H_2} is the partial pressure of hydrogen (atm); P_{O_2} is the partial pressure of oxygen (atm); F is the Faraday's constant $(= 9.648 \times 10^4$ C/mol); and T_{FC} is the fuel cell temperature (K).

By introducing the effects of double-layer capacitance charging at the electrode–electrolyte interface with a first-order dynamic, the following differential equation describes the activation overvoltage V_{act}:

$$\frac{dV_{act}}{dt} = \frac{I_d^{FC}}{C_{dl}} + \frac{I_d^{FC}V_{act}}{E_{act}C_{dl}} \tag{9.4}$$

where I_d^{FC} is the fuel cell current density in A/cm^2; C_{dl} is the double-layer capacitance (F); and E_{act} is the activation drop (V) defined by

$$E_{act} = \beta_1 + \beta_2 T_{FC} + \beta_3 T_{FC} \ln C_{o2} + \beta_4 T_{FC} \ln I_{FC} \tag{9.5}$$

Moreover, the parametric coefficients $\beta_1 - \beta_4$ can be determined as follows:

$$\beta_1 = -0.948$$
$$\beta_2 = -0.00286 + 0.0002 \ln A_{FC} + 4.3 \times 10^{-5} \ln C_{H_2} \qquad (9.6)$$
$$\beta_3 = 7.6 \times 10^{-5}$$
$$\beta_4 = -1.93 \times 10^{-4}$$

and

$$C_{O_2} = 1.97 \times 10^{-7} P_{O_2} \exp\left(\frac{498}{T_{FC}}\right)$$
$$C_{H_2} = 9.174 \times 10^{-7} P_{H_2} \exp\left(-\frac{77}{T_{FC}}\right) \qquad (9.7)$$

where A_{FC} is the effective fuel area (cm^2); C_{O_2} is the oxygen concentration at the cathode/membrane interface (mol/m^3); and C_{H_2} is the hydrogen concentration at the cathode/membrane interface (mol/m^3).

The ohmic overvoltage V_{act} is

$$V_{ohm} = \frac{l_M}{A_{FC}} r_M I_{FC} \qquad (9.8)$$

where l_M is the membrane thickness (cm) and r_M is the membrane resistivity (Ω), that is,

$$r_M = 181.6 \frac{1 + 0.3\dfrac{I_{FC}}{A_{FC}} + 0.062\left(\dfrac{T_{FC}}{303}\right)^2 \left(\dfrac{I_{FC}}{A_{FC}}\right)^{2.5}}{\left(11.866 - 3\dfrac{I_{FC}}{A_{FC}}\right)\exp\left(4.18\dfrac{T_{FC}-303}{T_{FC}}\right)} \qquad (9.9)$$

Based on the ideal gas law and the principle of mole conservation, the partial pressures of hydrogen (P_{H_2}) and oxygen (P_{O_2}) associated with reactant flow rates at the anode and cathode, along with the cell current, can be characterized, respectively, as follows:

$$\frac{dp_{H_2}}{dt} = \frac{RT_{FC}}{V_{anode}}\left[F_{H_2} - k_{anode}\left(p_{H_2} - p_{H_2,in}\right) - \frac{15}{F}I_{FC}\right] \qquad (9.10)$$

$$\frac{dp_{O_2}}{dt} = \frac{RT_{FC}}{V_{cathode}}\left[F_{O_2} - k_{cathode}\left(p_{O_2} - p_{BPR}\right) - \frac{7.5}{F}I_{FC}\right] \qquad (9.11)$$

where V_{anode} and $V_{cathode}$ denote the volumes of anode and cathode (m^3), respectively; F_{H_2} and F_{O_2} denote the molar flow rates (kmol/h) of hydrogen and oxygen, respectively; k_{anode} ($= 6.5 \times 10^{-2}$ mol/s·atm) and $k_{cathode}$ ($= 6.5 \times 10^{-2}$ mol/s·atm) represent

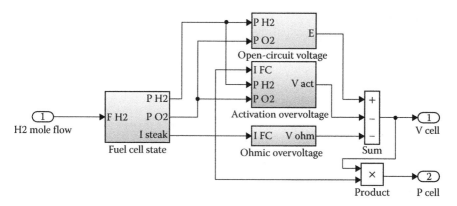

FIGURE 9.2   Model structure for PEM fuel cell.

the flow constants of anode and cathode, respectively; and P_{BPR} is the cell back pressure ($= 1$ atm).

Note that two pressure regulators with fixed parameters (i.e., k_{anode} and $k_{cathode}$) are added to manipulate the outlet hydrogen flow at the anode and the outlet oxygen flow at the cathode. Moreover, the voltage V_{stack} (V) and power P_{stack} (W) of the stack are given by

$$V_{stack} = 30V_{FC}$$
$$P_{FC} = V_{stack}I_{FC} \qquad\qquad (9.12)$$

These formulations have been adopted in this work to build a Simulink® code (see Figure 9.2) to simulate a PEMFC stack under the assumptions that the module is isothermal and its performance is only affected by the inlet flow rate of hydrogen (F_{H_2}).

9.2.3 WIND TURBINE

Dixon and Hall (2014) developed a generic model of the wind turbine, and they suggested that three design factors be considered:

1. Cut-in wind speed (v_{cut-in}): This is the lowest acceptable wind speed at which the turbine can deliver useful power. The value of v_{cut-in} is usually between 3 and 4 m/s.
2. Cut-out wind speed ($v_{cut-out}$): This is the maximum wind speed at which the turbine can operate safely. Such an upper bound is usually set according to the highest tolerable stresses of turbine components. When reaching this limit, the control system activates the braking system to bring the rotor to rest.

3. Rated output power (P_{rated}) and wind speed (v_{rated}): The electrical power output increases rapidly with wind speed, usually between 14 and 17m/s. At higher wind speeds, the power output remains constant at the rated value by the control system, making adjustment of the blade angles.

Notice first that the kinetic power P_{WT} (W) available in the wind is

$$P_{WT} = \frac{1}{2}\rho A_2 v^3 \tag{9.13}$$

where ρ is the air density (kg/m^3); A_2 is the disc area (m^2); and v is the velocity upstream of the disc (m/s). To mimic the cut-in and cut-out control in the wind turbine, the following logic constraints are imposed:

$$P_{WT}(t) = \begin{cases} 0 & v(t) \le v_{cut-in} \text{ or } v(t) \ge v_{cut-out} \\ \\ P_{WT} & v_{cut-in} < v(t) < v_{rated} \\ \\ P_{rated} & v_{rated} \le v(t) < v_{cut-out} \end{cases} \tag{9.14}$$

where P_{rated}, v_{cut-in}, and $v_{cut-out}$ are determined by the wind turbine supplier. The total power $P_{WTtotal}$ (W) when there are n turbines operating is

$$P_{WTtotal} = n \times P_{WT}(t) \tag{9.15}$$

Simulink® code is then built according to the previous model formulations for evaluating the power output of a wind turbine (see Figure 9.3).

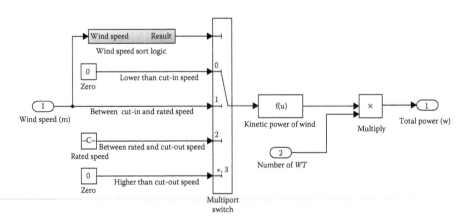

FIGURE 9.3   Model structure for a wind turbine.

9.2.4 ALKALINE ELECTROLYZER

An existing mathematical model of the alkaline electrolyzer (Ulleberg et al., 2010) is adopted here. The empirical current–voltage (I – U) relationships are used to model the electrode kinetics of an electrolyzer. In electrolysis, the operating cell voltage is the aggregate of reversible overvoltage, activation overvoltage, and ohmic overvoltage. Thus, the operating cell voltage is equal to:

$$V_{cell} = V_{rev} + V_{act} + V_{ohm} \qquad (9.16)$$

Taking account of temperature dependency for activation and ohmic overvoltage, the empirical current–voltage (I – U) can be expressed as:

$$V = V_{rev} + s \ \ln\left[\frac{I}{A_E}\left(t_1 + \frac{t_2}{T_{ELY}} + \frac{t_3}{T_{ELY}^2}\right) + 1\right] + \frac{1}{A_E}(r_1 + r_2 T_{ELY}) \qquad (9.17)$$

where T_{ELY} denotes the cell temperature (°C); A_E denotes the area of the electrode (m^2); I denotes the electrical current flow through the cell; s ($= 0.185$ V), t_1 ($= -0.1804$ $A^{-1}m^2$), t_2 ($= 18$ $A^{-1}m^2$°C), and t_3 ($= -0.1804$ $A^{-1}m^2$°C^2) are the coefficients for activation overvoltage on electrodes; and r_1 ($= 8.05 \times 10^{-5}$ Ωm^2) and r_2 ($= -2.5 \times 10^{-7}$ Ωm^2°C^{-1}) are the coefficients for the ohmic resistance of the electrolyte.

Based on Faraday's law, the production of hydrogen is affected by the electrical current in the external circuit. Moreover, the theoretical amount of the total hydrogen production rate in the electrolyzer should be multiplied by the Faraday efficiency to gain the actual amount of the hydrogen production rate. The equation for the operation is

$$\dot{n}H_2 = \eta_F \frac{n_c I}{ZF} \qquad (9.18)$$

where n_c represents the number of cells in series; Z denotes the electrons exchanged during the electrolysis process; and F denotes the Faraday constant. The Faraday efficiency (η_F) is caused by parasitic current losses in which it will increase due to the decrease in current densities. Because resistance decreases as temperature rises, the net effect is a lowered Faraday efficiency due to an increased parasitic current loss. This phenomenon can be described as follows:

$$\eta_F = \frac{\left(\frac{I}{A}\right)^2}{f_1 + \left(\frac{I}{A}\right)^2} f_2 \qquad (9.19)$$

where f_1 ($= 250$ mA^2cm^{-4}) and f_2 ($= 0.98$) are Faraday efficiency parameters.

A Simulink® code can also be built using these model formulations to evaluate the hydrogen production rate of the alkaline electrolyzer (see Figure 9.4).

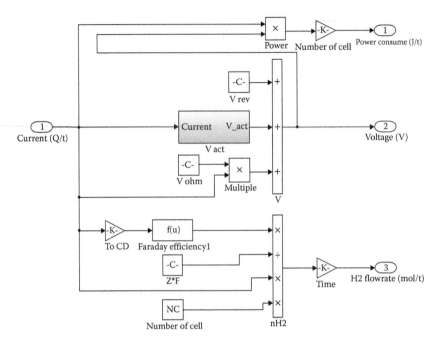

FIGURE 9.4   Model structure for alkaline electrolyzer.

9.2.5   Hydrogen Storage Tank

The high-pressure hydrogen storage tank is modeled by the ideal gas law. The pressure change in the tank can be expressed as:

$$P_{tank} = \frac{nRT_{tank}}{V_{tank}} \tag{9.20}$$

where R ($= 8.314 \times 10^{-2}$ L bar K^{-1} mol^{-1}) is the ideal gas constant; T_{tank} is the temperature of the hydrogen gas (K); and n is the amount of hydrogen (mol). This model assumes that the atoms in an ideal gas behave as rigid spheres and the collisions are perfectly elastic. Also, there are no intermolecular attractive forces between molecules. In an ideal gas, all the internal energy is present in the form of kinetic energy, and the change of internal energy is accompanied by a temperature change.

A Simulink® block can be built with the previous formulations to describe the pressure variation in the storage tank (see Figure 9.5).

9.2.6   Battery

The discharge model of the Li-ion battery in MATLAB (MathWorks, 2016) accurately depicts the voltage dynamics when the current varies. The expression of battery voltage is:

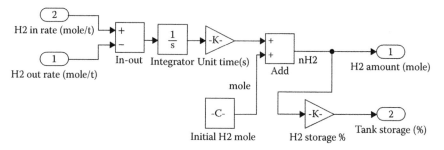

FIGURE 9.5 Model structure for the hydrogen storage tank.

$$V_{batt} = E_0 - V_{pol}\ \text{it} - R_i I_{batt} + A\exp(-B\cdot\text{it}) - R_{pol}I^*_{batt} \tag{9.21}$$

where V_{batt} denotes the battery voltage (V); E_0 is a reference constant (V); $V_{pol} = K_V \dfrac{Q}{Q-it}$ is the polarization voltage (V); $R_{pol} = K_R \dfrac{Q}{Q-it}$ is the polarization resistance (Ω); K_V (Ω) and K_R (Ω) denote the polarization voltage and resistance constants, respectively; Q is the battery capacity (A h); A is the exponential zone amplitude (V); B is the inverse of the zone time constant (A); R_i is the internal resistance (Ω); I_{batt} denotes the battery current; I^*_{batt} denotes the filtered current, which was determined using the first-order low-pass filter; and it $= \int I_{batt}dt$ represents the actual battery charge or

$$\frac{d\ \text{it}}{dt} = I_{batt} \tag{9.22}$$

On the other hand, note that the voltage increases rapidly when the Li-ion batteries reach the full charge. A polarization resistance term can model this phenomenon. In the charge mode, the polarization resistance increases until the battery is almost full (it $\approx$ 0). Above this point, the polarization resistance should rise abruptly. Instead of the polarization resistance in the aforementioned discharge model (i.e., see Equation 9.21), this resistance should be

$$R_{pol} = K_R \frac{Q}{\text{it}} \tag{9.23}$$

Note that as it $\rightarrow$ 0, it clearly should approach infinity. However, this is not exactly the case in practice. Experimental results have shown that the contribution of the polarization resistance shifts about 10% of the capacity of the battery. In the charging model, instead of the last term on the right side of Equation 9.21, the polarization resistance should be

$$R_{pol,charge} = K_R \frac{Q}{\text{it} - 0.1Q} \tag{9.24}$$

More specifically, the charging model used in the present study is

$$V_{batt} = E_0 - V_{pol} \, \text{it} - R_i I_{batt} + A \exp(-B \cdot \text{it}) - R_{pol,charge} I_{batt}^* \qquad (9.25)$$

The corresponding *state of charge* (SOC) is

$$SOC = 100\left(1 - \frac{\text{it}}{Q}\right) \qquad (9.26)$$

In the following case studies, the simulation code for the Li-ion battery is taken from the Simulink® model library. For further understanding, the model structure and parameter settings can be found at http://www.mathworks.com/help/physmod/ sps/powersys/ref/battery.html. Let us also assume that the battery connects directly to a DC bus to store excess power during the daytime and provides electricity at night when there is a shortage.

9.3   SIMULATION-BASED VERTEX METHOD

Although other alternative numerical strategies (such as the time-dependent version of the active set method described in the previous chapters) may be utilized as well, a simulation-based optimization method has been adopted here to compute the flexibility indices. There are several advantages in taking such an approach. First of all, to relieve the heavy computation load caused by the overwhelmingly large number of possible candidate solutions in applying the conventional vertex method, some effective heuristics can be easily incorporated into the search procedure to promote convergence. This computation strategy is also believed to be especially attractive in assessing the operational flexibility of a complex process in which a large variety of different units are integrated to achieve a common objective. By incorporating the independently developed and individually modularized simulation codes for all embedded components into a generalized framework, any system configuration can be analyzed and evaluated efficiently.

9.3.1   VERTEX SELECTION HEURISTICS

The dynamic version of the traditional vertex method can be formulated as the two-level optimization problem described in section 4.2.2, that is,

$$FI_d = \min_k \max_{z(t), x(t)} \delta_d \qquad (9.27)$$

subject to Equations 4.1 and 4.2 and the following constraint in a functional space formed by all possible time profiles of $\theta(t)$:

$$\theta(t) = \theta^k(t) = \theta^N(t) + \delta_d \Delta \, \theta^k(t) \qquad (9.28)$$

where $\Delta\theta^k(t)$ denotes a vector pointing from the nominal point $\theta^N(t)$ toward the kth vertex ($k = 1,2,3,\ldots,2^{N_p}$ and N_p is the number of uncertain parameters) at time t.

Note that each element in vector $\Delta\Theta^k(t)$ should be obtained from the corresponding entry in either $-\Delta\Theta^-(t)$ or $\Delta\Theta^+(t)$. Because the total number of vertices increases exponentially with the number of uncertain parameters, there is a need to reduce at least some of the search effort. For this purpose, the physical insights of the given system may be utilized to facilitate elimination of a portion of vertices. Intuitively speaking, the critical deviations in uncertain parameters from their nominal levels should drive the system toward one or more boundaries of the feasible region. These critical deviations can only be determined on a case-by-case basis in particular applications.

On the other hand, the vertex directions of a hypercube defined by Equation 5.5 can be expressed mathematically as

$$\Theta(H) = \delta_t \Delta \; \Theta^k \tag{9.29}$$

where $\Delta\Theta^k$ denotes a vector pointing from the origin (i.e., the nominal point) in the N_p-dimensional Euclidean space toward the kth vertex, and each element in $\Delta\Theta^k$ must be the same as the corresponding entry in either $-\Delta\Theta^-$ or $\Delta\Theta^+$. From the definition of

$\Theta(H)\left(= \int\limits_0^H \left[\Theta(\tau) - \Theta^N(\tau)\right] d\tau\right)$, it is clear that every vertex can be reached with an

infinite number of time profiles that are constrained by Equation 5.4. Therefore, to be able to implement the temporal version of the extended vertex method in realistic applications, it is obviously necessary to reduce the search space to a manageable size. It has been found that in addition to Equation 5.4, an extra heuristic should be adopted in computing FI_t to further constrain the candidate time profiles of the uncertain parameters for use in Equation 5.5, that is,

$$\Delta\hat{\theta}_n^k\left(t; t_0^n, t_f^n\right) = \hat{\theta}_n^k\left(t; t_0^n, t_f^n\right) - \theta_n^N(t) = \begin{cases} \Delta\theta_n^k(t), & \text{if } 0 < t_0^n \le t \le t_f^n < H \\ 0, & \text{if } 0 \le t < t_0^n \text{ or } t_f^n < t \le H \end{cases} \tag{9.30}$$

where $n = 1, 2, \ldots, N_p$; $k = 1, 2, 3, \ldots, 2^{N_p}$; $\theta_n^N(t)$ is the nth element of vector $\theta^N(t)$ defined in Equation 5.5 and $\Delta\theta_n^k(t)$ in time interval $\left[t_0^n, t_f^n\right]$ represents the nth element of a vector pointing from the nominal point $\theta^N(t)$ toward the kth vertex of the functional hypercube defined by Equation 5.5. More specifically, the position of this vertex can be expressed as

$$\theta(t) = \hat{\theta}^k(t) = \theta^N(t) + \Delta \; \hat{\theta}^k(t)$$

where

$$\hat{\theta}^k(t) = \left[\hat{\theta}_1^k\left(t; t_0^1, t_f^1\right) \quad \hat{\theta}_2^k\left(t; t_0^2, t_f^2\right) \quad \cdots \quad \hat{\theta}_{N_p}^k\left(t; t_0^{N_p}, t_f^{N_p}\right)\right]^T \tag{9.31}$$

$$\theta^N(t) = \left[\theta_1^N(t) \quad \theta_2^N(t) \quad \cdots \quad \theta_{N_p}^N(t)\right]^T$$

$$\Delta\hat{\theta}^k(t) = \left[\Delta\hat{\theta}_1^k\left(t;t_0^1,t_f^1\right) \quad \Delta\hat{\theta}_2^k\left(t;t_0^2,t_f^2\right) \quad \cdots \quad \Delta\hat{\theta}_{N_p}^k\left(t;t_0^{N_p},t_f^{N_p}\right) \right]^T$$

As mentioned previously, $\Delta\hat{\theta}^k(t)$ can be viewed as a vector in the *functional space* of $\theta(t)$, which starts from the nominal point $\theta^N(t)$ and ends at the kth vertex only in time interval $\left[t_0^n, t_f^n\right]$ and, in addition, each element of $\Delta\hat{\theta}^k(t)$ should be obtained from the corresponding entry in either $-\Delta\theta^-(t)$ or $\Delta\theta^+(t)$. Notice also that as clearly indicated in Equation 9.31, the allowed deviation in each uncertain parameter may begin and terminate at instances that are not the same as those of the other parameters. Thus, FI_t can, in principle, be determined as follows:

$$FI_t = \min_k \ \max_{\mathbf{z}(t),\mathbf{x}(t)} \ \delta_t \tag{9.32}$$

subject to Equations 4.1 through 4.3 and 9.29 through 9.31.

Finally, although the justification for these heuristics is derived from an intuitive belief—that is, the most severe disturbance a realistic process can withstand is usually the one with the largest possible magnitude in the shortest period of time—its validity has been verified by numerically simulating the worst-case scenarios in extensive case studies and by solving the same problems independently with the extended active set method.

9.3.2 SIMULATION-BASED ALGORITHM FOR COMPUTING DYNAMIC FLEXIBILITY INDICES

The dynamic flexibility index is determined by Equations 4.1, 4.2, 9.27, and 9.28. In the proposed computation procedure, the bisection strategy is utilized to search for the maximum value of δ_d under the constraint of Equations 4.1, 4.2, and 9.28. The detailed solution algorithm can be illustrated with the flowchart in Figure 9.6. Notice that there are three components in this computation procedure:

1. The maximum δ_d corresponding to every candidate vertex should be computed, and the dynamic flexibility index is the smallest among them.
2. Before starting the bisection search, the initial range of δ, that is, $\left[\delta_0^{\min}, \delta_0^{\max}\right]$, should be fixed *a priori*. $\delta_0^{\min}$ is usually set at 0, whereas $\delta_0^{\max}$ may assume the value of 3 just to save the computation time (a larger value can be chosen if a higher flexibility is anticipated). At iteration $nitr$, the midpoint of $\left[\delta_{nitr}^{\min}, \delta_{nitr}^{\max}\right]$ is adopted for generating the maximum and/ or minimum uncertain parameter profiles. If the postulated profiles fail the feasibility check after simlation, set the current δ^k as the new upper bound $\delta_{nitr+1}^{\max}$. Otherwise, replace the lower bound with the current δ^k, that is, set it to be $\delta_{nitr+1}^{\min}$. If δ_d is indeed present in the initial interval

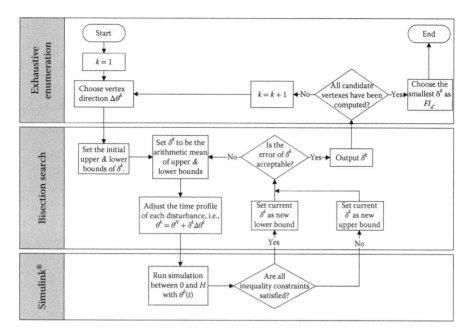

FIGURE 9.6 Optimization algorithm for computing the dynamic flexibility index.

$\left[\delta_0^{min}, \delta_0^{max}\right]$, the termination criterion after *nstep* iterations can obviously be expressed as

$$\frac{\delta_0^{max} - \delta_0^{min}}{2^{nitr}} < \varepsilon \tag{9.33}$$

3. This step performs the feasibility check according to Equation 4.2. Specifically, the system behavior is simulated according to Equations 4.1 and 9.28 and a given value of δ^k.

9.3.3 SIMULATION-BASED ALGORITHM FOR COMPUTING TEMPORAL FLEXIBILITY INDICES

As mentioned previously, the temporal flexibility index can be determined by Equations 4.1 through 4.3 and 9.29 through 9.32. There are three groups of decision variables in this problem, respectively, characterizing the vertex direction $\Delta\Theta^k$, the scalar variable δ^k, and the initial and final times of each parameter deviation, that is, t_0^n and t_f^n in Equation 9.30. Given a vertex direction $\Delta\Theta^k$, the subgoal of optimization is to find the largest *feasible* δ^k in all possible scenarios defined by Equation 9.30. The genetic algorithm (GA) is chosen here to search for the optimal t_0^ns, whereas the bisection strategy is used to identify the maximum value of δ^k according to a set of given initial times. Note that it is only necessary to consider the candidate vertices (which are selected heuristically), and the smallest objective value should be chosen as FI_t. Figure 9.7 shows the corresponding flowchart. Notice that there are also three components in this procedure:

1. For a candidate vertex, the evolution procedure of the GA starts with a group of initial solution candidates. To calculate the temporal flexibility index, each chromosome is made up of an array of starting times defined by Equation 9.30, that is, t_0^n and $n = 1, 2, \ldots, N_p$. The chosen population size is 80, and each chromosome is made of three random integer numbers between 0 and 240, which represent the three initial disturbance times in multiples of 0.1 hour. The fitness measure is the value of δ_t determined according to Equation 9.32 (which can be implemented via a bisection search), and a portion of the population is then selected with the roulette wheel. The next step creates a random binary array to perform crossover between two chromosomes. For example, if the two selected parents are

$$\text{parent } 1 = \begin{bmatrix} w & x & y & z \end{bmatrix}$$
$$\text{parent } 2 = \begin{bmatrix} a & b & c & d \end{bmatrix}$$
(9.34)

and the random binary array is [1 0 1 1], the resulting chromosomes should be [w b y z]. The mutation step adds a random number taken from a Gaussian distribution with a mean 0 to each parent. The evolution process is stopped either after 20 generations or when the average relative change in the best fitness value is smaller than 0.003. Finally, notice that it is necessary to check the candidate vertices exhaustively.

2. With the given initial disturbance times (t_0^n s), the bisection search is used to identify the largest δ^k that satisfies all inequality constraints defined in Equations 4.2 and 5.5. First, select an initial interval $\left[\delta_0^{\min}, \delta_0^{\max} \right]$ that contains δ_t. The methods to set $\delta_0^{\min}$, $\delta_0^{\max}$, and the termination condition are the same as those for computing the dynamic flexibility. However, the algorithm applies the midpoint of $\left[\delta_n^{\min}, \delta_n^{\max} \right]$ into Equation 9.29 to define the amount of accumulated deviation. The next step is to check if one or more inequality constraints become invalid. If yes, the current midpoint should replace $\delta_n^{\max}$. If otherwise, the current midpoint should replace $\delta_n^{\min}$.

3. After fixing the initial disturbance times t_0^n and δ^k, time profiles of uncertain parameters $\hat{\theta}^k(t)$ can be determined by Equations 9.29 through 9.31. The Simulink® codes mentioned previously in Section 2 then start to simulate the hybrid power system. As soon as the simulation time reaches t_0^n, the nth uncertain parameter changes to $\theta_n^N(t) + \Delta\hat{\theta}_n^k(t)$. When the accumulated deviation of the nth uncertain parameter reaches $\delta^k \Delta\Theta_n^k$, the nth uncertain parameter changes back to $\theta_n^N(t)$. Finally, the feasibility test of each inequality constraint is required after completing the simulation run.

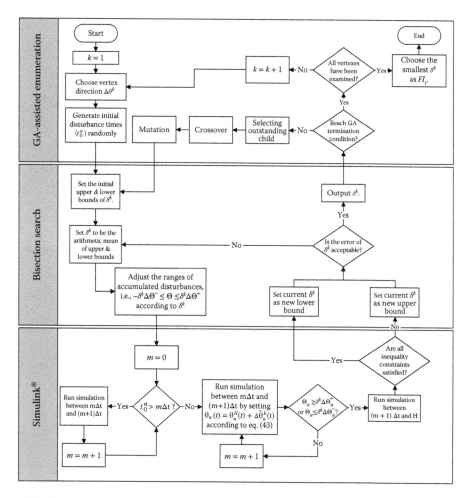

FIGURE 9.7 Optimization algorithm for computing the temporal flexibility index.

9.3.4 A Benchmark Example

To better illustrate the proposed simulation-based optimization strategies, let us revisit the simple buffer vessel described previously in Section 4.3 (see Figure 4.3). As mentioned before, the dynamic model of this system can be formulated as follows:

$$A_{tank}\frac{dh}{dt} = \theta(t) - k\sqrt{h} \tag{9.35}$$

In this model, h stands for the height of the liquid level (m); θ denotes the feed flow rate ($m^3 min^{-1}$); A_{tank} ($5m^2$) is the cross-section area of the tank; and k ($= \sqrt{5}/10\ m^{5/2}min^{-1}$) is a proportional constant. A Simulink® code has been built from the model formulations to simulate the dynamic behavior of this buffer system (see Figure 9.8). Specifically, this code integrates the difference between input and output flow rates to determine the transient variations of liquid volume and level. For illustration purposes, the following assumptions are adopted:

- The height of the tank is 10 m and the minimum required liquid level is 1 m.
- The time horizon is between 0 and 800 (i.e., $0 \leq t < 800$).
- The feed flow rate is the only uncertain parameter in this example, and its nominal and anticipated positive and negative deviation profiles are shown in Figure 9.9.
- The maximum accumulated positive and negative deviations in the feed are chosen to be 20 m^3 ($\Delta\Theta^- = \Delta\Theta^+ = 20$).
- The initial height of the liquid level is 5 m.

For the simulation-based flexibility analyses, the equality constraint is Equation 9.35, and the inequality constraints should be written as

$$1 \leq h(t) \leq 10 \tag{9.36}$$

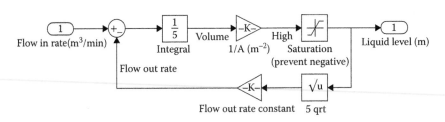

FIGURE 9.8 Model structure of the liquid buffer tank.

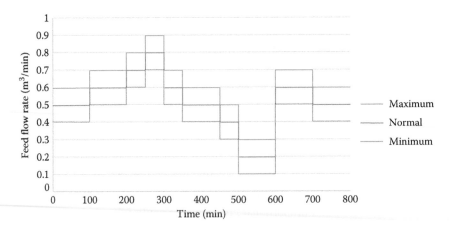

FIGURE 9.9 Nominal feed rate and its upper and lower limits in a single period.

Let us first try to identify the candidate vertices. Obviously, it can be expected that the deviations in the feed flow rate may drive the liquid level toward the upper or lower bounds. Numerical simulation results (see Figure 9.10) also confirmed these predictions. Hence, both vertices should be taken into account in flexibility analyses.

To evaluate the dynamic flexibility index, it is necessary first to fix two algorithmic parameters. The initial upper bound δ_0^{max} is set to be 3, whereas the termination upper bound ϵ is 0.001. The resulting dynamic flexibility index was found to be 0.3636.

The corresponding simulation results in Figure 9.11 show that at $\delta_d = 0.3636$, the liquid level always satisfies the constraints in Equation 9.36 and just touches the lower limit at 600 min. Thus, the tank system can withstand any uncertain feed rate profile by reducing the permissible range of the uncertain parameter, that is,

$$\theta^N(t) - 0.3636 \times \Delta\theta^-(t) \leq \theta(t) \leq \theta^N(t) + 0.3636 \times \Delta\theta^+(t) \qquad (9.37)$$

where $\theta^N(t)$, $\Delta\theta^-(t)$ and $\Delta\theta^+(t)$ are defined in Figure 9.9.

To calculate the temporal flexibility index, it is also necessary to fix two algorithmic parameters first. The initial δ_0^{max} is set to be 2 to reduce the calculation time, and the evolution process of the GA is terminated either after 20 generations or the average relative change in the best fitness value is smaller than 0.003. The proposed optimization runs were repeated three times according to the algorithm given in Figure 9.7. Starting with 480 sets of initial disturbance times, each run took six generations to reach the terminal condition, and the average computation time is 234.1 seconds on an Intel® Core(TM) i7-4790 CPU @ 3.60 GHz. The temporal flexibility index was found to be 0.1836 for which the disturbance exists in the time interval between 562.0 and 598.7 minutes. One can observe from the simulation results in Figure 9.12 that as the feed rate drops to its minimum level at 562 minutes and stays at that level until the instance when the feed decrease results in an accumulated shortage of 3.672 ($= 20 \times 0.1836$) m³, the liquid level just touches the lower limit of 1 m.

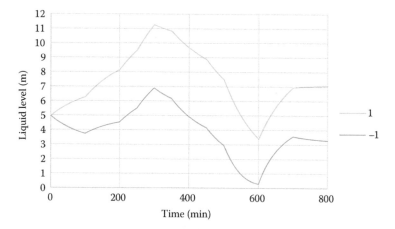

FIGURE 9.10 Liquid level response to flow rate at upper limit and lower limit.

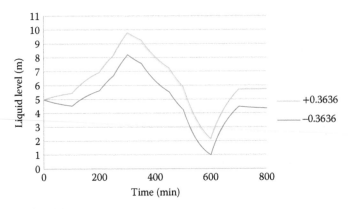

FIGURE 9.11 Liquid level response to different δ_d.

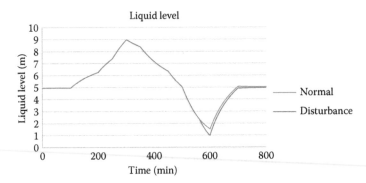

FIGURE 9.12 Simulation results of the worst-case scenarios ($FI_t = 0.1836$, $A = 5\text{m}^2$, $t_0 = 562$ min, $t_f = 598.7$ min).

TABLE 9.1
Computation Times of Two Alternative Vertex Methods

	Dynamic Flexibility Index		Temporal Flexibility Index	
	Simulation-based vertex method	Extended vertex method	Simulation-based vertex method	Extended vertex method
Time consumed (s)	0.93	0.91	234.1	394

In Table 9.1, the computation performance of the current method is compared with that of the vertex method developed by Kuo and Chang (2016). Note that although the proposed strategy does not have a significant improvement in accuracy, a reduction of 40% computation time for the temporal flexibility index can be achieved due to the use of vertex selection heuristics and the GA.

Based on these dynamic and temporal flexibility indices, one could conclude that the buffer tank is not well designed to withstand the anticipated uncertain disturbances. The design targets $FI_d = 1$ and $FI_t = 1$ can be achieved by enlarging the cross-sectional area of the buffer tank to 8.25 m² and 7.02 m², respectively. This difference is due primarily to their different definitions. The former considers the long-term disturbances throughout the entire horizon, whereas the latter cares only about the accumulated effects of temporary disturbances.

9.4 CASE STUDIES

The case studies presented here are used to demonstrate the important roles of flexibility analyses in evaluating the operational performance of a given hybrid power system for standalone applications. To fix the ideas, let us consider the preliminary designs aimed at providing uninterrupted power for the Annan campus of National Cheng Kung University in Tainan, Taiwan. The optimal system configurations are determined primarily by minimizing the total annual cost (TAC_{system}) under the condition that $FI_d = 1$ and/or $FI_t = 1$.

9.4.1 MODEL PARAMETERS

Let us consider a simplified schematic of the hybrid power plant in Figure 9.13. To facilitate flexibility analyses, the unit models have been established according to the published literature and vendor's manuals. The model parameters of PV and fuel cells have been taken, respectively, from the datasheet of BP Solar's MSX 60 and from Wu et al. (2013), where each unit was sized to be equivalent to the 7-kW Panasonic household model "ENE-FARM," whereas those of wind turbine have been adopted from the design specifications of Hi-VAWT's DS-700 Vertical Axis Wind Turbines. The battery parameters were extracted from the MATLAB/Simulink Simscape block, and the size of each module was chosen to be that of a 10-kWh Tesla Powerwall manufactured by Tesla Motors. For simplicity, an ideal DC-AC inverter is assumed to be available to provide the 60 Hz electricity, and the efficiency is 100% (Chayawatto et al., 2009). It is also assumed that all components

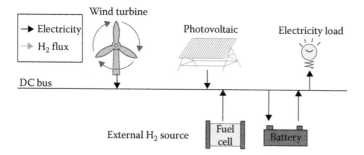

FIGURE 9.13 Simplified schematics of PV/FC/WT hybrid power generation system.

are interconnected via an electrical bus and, thus, the charging/discharging behavior of the battery can be described as follows:

$$I_{batt} = I_{PV} + I_{FC} + I_{WT} - I_{demand} \tag{9.38}$$

The hydrogen feed rate was set to be a constant in this case at 0.07 kmol/h, and a total of three uncertain parameters have been considered in the case studies: the wind speed, solar irradiance, and power demand. The time profiles of solar irradiance and wind speed at Annan are presented in Figures 9.14 and 9.15, respectively, and the time profiles of $\theta^N(t)$, $\Delta\theta^-(t)$, and $\Delta\theta^+(t)$ in Equation 4.3 can then be estimated accordingly. On the other hand, the average daily power demand and its upper and lower bounds on the Annan campus were estimated to be 719 kWh, 862.8 kWh, and 575.2 kWh, respectively. The corresponding hourly power demands in a day (see Figure 9.16) were established by assuming that these time profiles are all identical to that of a typical Taiwanese office building. The second and third columns of Table 9.2 list the anticipated values of $\Delta\Theta^-$ and $\Delta\Theta^+$ for each uncertain parameter, respectively. By comparing this with the integral values of the deviations taken from Figures 9.14 through 9.16, respectively, in the last two columns, one can observe that the inequalities given in Equations 4.21 and 4.42 are satisfied.

9.4.2 DESIGN VARIABLES

As a design variable, the energy supply-to-demand ratio should be specified *a priori*, that is,

$$r_{SD} = \frac{\text{Nominal daily energy supply}}{\text{Nominal daily energy demand}} \tag{9.39}$$

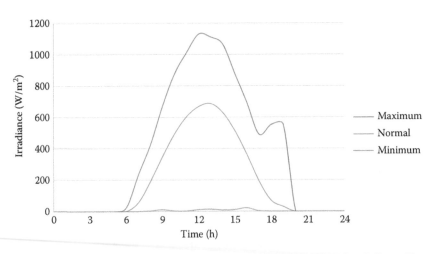

FIGURE 9.14 Hourly sun irradiance profile of Annan campus of National Cheng Kung University in Tainan, Taiwan.

Its denominator can be computed by numerically integrating the power demand profile in Figure 9.16 over a period of 24 hours, whereas the numerator is computed according to the following formula:

Nominal daily energy supply =

$$E_{PV}^{N} \times N_{PV} + E_{FC}^{N} \times N_{FC} + E_{WT}^{N} \times N_{WT} + E_{B}^{I} \times N_{B}$$ (9.40)

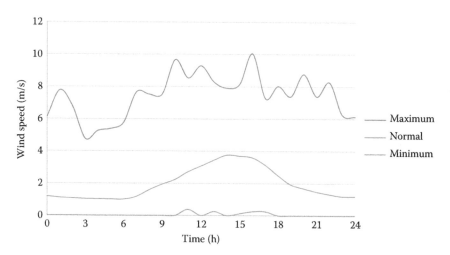

FIGURE 9.15 Hourly wind speed profile of Annan campus of National Cheng Kung University in Tainan, Taiwan.

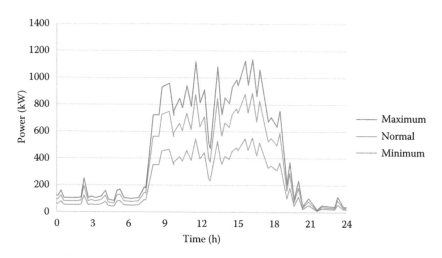

FIGURE 9.16 Assumed hourly power consumption profile of Annan campus of National Cheng Kung University in Tainan, Taiwan.

TABLE 9.2

Accumulated Deviations of Three Uncertain Parameters in Case Studies

	$\Delta\Theta^+$	$\Delta\Theta^-$	$\int_0^H \Delta\theta^+(\tau)d\tau$	$\int_0^H \Delta\theta^-(\tau)d\tau$
Wind speed (m hr/s)	57.7	28.2	131.8	47.4
Solar irradiance (W hr/m²)	3879.1	4557.6	4877.8	4800.0
Power consumption (kWh)	1896.0	2575.5	2404.2	3265.9

where E_{PV}^N, E_{FC}^N, and E_{WT}^N denote the nominal daily energy outputs from PV, FC, and WT modules, respectively, and they can be determined by integrating the nominal profiles of the output powers produced by the corresponding modules with MATLAB/Simulink; E_B^I denotes the initial energy stored in the battery, which can be viewed as a design specification; and N_{PV}, N_{FC}, N_{WT}, and N_B denote the numbers of corresponding modules.

The four corresponding energy ratios are as follows:

$$r_{PV} = \frac{E_{PV}^N \times N_{PV}}{\text{Nominal daily energy supply}} \tag{9.41}$$

$$r_{FC} = \frac{E_{FC}^N \times N_{FC}}{\text{Nominal daily energy supply}} \tag{9.42}$$

$$r_{WT} = \frac{E_{WT}^N \times N_{WT}}{\text{Nominal daily energy supply}} \tag{9.43}$$

$$r_B = \frac{E_B^N \times N_B}{\text{Nominal daily energy supply}} \tag{9.44}$$

$$r_{PV} + r_{FC} + r_{WT} + r_B = 1 \tag{9.45}$$

9.4.3 PERFORMANCE MEASURES

For the purpose of quantitatively evaluating the economic performance, the capital cost of every component in the hybrid system was estimated according to literature data. Based on the unit cost reported in Amazon.com in 2015, a figure of $346 was adopted as the approximate cost of a BP Solar's MSX 60 module. It was also assumed that a 7.5 kW Panasonic household fuel cell (ENE-FARM) system could be purchased at the price of $226,000 in 2015. The installation cost of a Hi-VAWT's DS-700 VAWT was set at $1000. Finally, a purchasing cost of $3,500 was used as the capital cost of a 10 kWh Tesla Powerwall of Tesla Motors in 2015. To calculate the total annualized cost (TAC_{system}), an annual interest rate of 5% is adopted in this study, and the lifetime of each component is set to be 5 years. The annualized capital cost is determined as follows (Nelson et al., 2006):

$$AC_{unit} = C_{unit}^{in} \frac{i_r(1+i_r)^l}{(1+i_r)^l - 1}$$
(9.46)

where AC_{unit} is the annualized capital cost of any unit in the hybrid power system; C_{unit}^{in} is the purchase cost of the same unit; i_r is the interest rate; and l is the expected lifetime. The total annual cost of a unit (TAC_{unit}) can then be calculated as follows:

$$TAC_{unit} = AC_{unit} + OC_{unit}$$
(9.47)

where OC_{unit} is the total annual operating and/or maintenance cost of the corresponding unit. Thus, the total annual cost of the hybrid power system can be expressed as

$$TAC_{system} = TAC_{PV} + TAC_{FC} + TAC_{WT} + TAC_B$$
(9.48)

It was also assumed that hydrogen can be purchased and continuously delivered at a cost of 1.25 \$/kmol (Thomas et al., 1998), which should be included in the operating cost of the fuel cell, and the maintenance cost of the WT unit was set at 3% of its capital cost (Nelson et al., 2006).

To identify the optimal system configuration, the values of FI_d for $r_{SD} = 2$ were determined by varying all the possible energy ratios of PV, FC, WT, and battery modules according to the same initial SOC for the battery bank (50%). As mentioned before, a FI_d value of 1 is targeted to ensure feasible operation throughout the entire time horizon under the expected uncertain disturbances. This target surface is given in Figure 9.17, and the colors on surface map different TAC_{system} s. The numbers of PV, FC, WT, and battery modules in each configuration on the target surface, the corresponding energy ratios, and TAC_{system} are all listed in Table 9.3.

It can be observed from Table 9.3 that, due to the stable power supply of the fuel cell, the FC energy ratio is often quite large in the cheaper designs. On the other hand, the PV and WT modules are not really in demand, as their energy supplies cannot be confirmed (see Figures 9.14 and 9.15). To achieve the flexibility target, a large number of PV units are usually needed so as to keep the minimum achievable power output at a high enough level. Notice also that if required in a feasible configuration, the number of WT modules tends to be quite large. This is because the wind speed on the Annan campus is often too low to generate power economically (Figure 9.15). Consequently, the hybrid system with the lowest TAC_{system} on the target surface (\$693,418.34) consists of only three types of components: 80 FC modules, 143 PV modules, and 310 battery modules.

The target surface for $FI_t = 1$ is presented in Figure 9.18. The numbers of PV, FC, WT, and battery modules in each configuration on the target surface, the

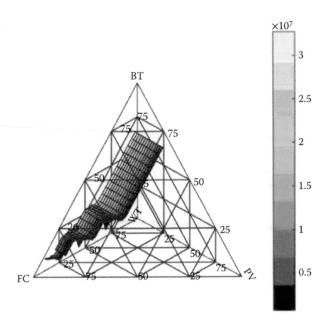

FIGURE 9.17 Surface area that represents all the possible ratios of PV, FC, WT, and battery modules with FI_d value of 1 for $r_{SD} = 2$.

corresponding energy ratios, and TAC_{system} are shown in Table 9.4. It was found that the optimal configuration in this case is actually the same as that for $FI_d = 1$. Notice that the feasible region is located above the colored surface in either Figure 9.17 or 9.18. The volume of the feasible region corresponding to $FI_t = 1$ (see Figure 9.18) is clearly larger than that for $FI_d = 1$ (see Figure 9.17). This is due to the chosen range for the accumulated deviations in Table 9.2. It should also be noted that the feasible volume is expanded more toward the PV apex but not WT. This is because the selected deviation of solar irradiance has the ability to generate power, whereas the selected wind speed in Annan campus is still smaller than the cut-in wind speed. The aforementioned surface positions reveal that $FI_d = 1$ is a more stringent design criterion and, consequently, FI_d should be computed first. If the budget to achieve $FI_d = 1$ is prohibiting, one can then switch to a relaxed criterion, that is, $FI_t = 1$, to identify an economically and operationally feasible system configuration.

From the previous result, we can observe that PV and WT units are not favored for the Annan campus of National Cheng Kung University in Tainan. This conclusion is probably due to the inherent natures of both energy resources and performance metrics. First of all, the wind and solar power sources are clearly not rich enough in Annan for producing economically attractive power. On the other hand, a flexibility analysis always considers the worst-case scenario, which is associated with the lower bounds of wind speed and sun irradiance at all times. Thus, if one wants to make use of a larger portion of the wind or solar energy in the hybrid power system, one must take the risk of relaxing the constraint of $FI_d = 1$. To fix the idea,

TABLE 9.3

Number of Units and Power Ratios Required for $FI_d = 1$ and $r_{SD} = 2$

N_{PV}	N_{FC}	N_{WT}	N_B	r_{PV}	r_{FC}	r_{WT}	r_B	TAC ($)
0	90	0	310	0	0.91	0	0.09	$714,332.85
143	80	0	310	0.1	0.81	0	0.09	**$693,418.34**
0	80	156555	310	0	0.81	0.1	0.09	$11,656,429.15
143	71	0	654	0.1	0.71	0	0.19	$883,844.12
0	71	156555	654	0	0.71	0.1	0.19	$11,846,854.93
287	61	0	654	0.2	0.61	0	0.19	$862,929.62
143	61	156555	654	0.1	0.61	0.1	0.19	$11,825,940.43
0	61	313109	654	0	0.61	0.2	0.19	$22,788,951.24
287	51	0	999	0.2	0.51	0	0.29	$1,053,355.39
143	51	156555	999	0.1	0.51	0.1	0.29	$12,016,366.21
0	51	313109	999	0	0.51	0.2	0.29	$22,979,377.02
430	41	0	999	0.3	0.41	0	0.29	$1,032,440.89
287	41	156555	999	0.2	0.41	0.1	0.29	$11,995,451.70
143	41	313109	999	0.1	0.41	0.2	0.29	$22,958,462.51
0	41	469664	999	0	0.41	0.3	0.29	$33,921,473.33
430	31	0	1343	0.3	0.31	0	0.39	$1,222,866.67
287	31	156555	1343	0.2	0.31	0.1	0.39	$12,185,877.48
143	31	313109	1343	0.1	0.31	0.2	0.39	$23,148,888.29
0	31	469664	1343	0	0.31	0.3	0.39	$34,111,899.10
430	21	0	1688	0.3	0.21	0	0.49	$1,413,292.45
287	21	156555	1688	0.2	0.21	0.1	0.49	$12,376,303.26
143	21	313109	1688	0.1	0.21	0.2	0.49	$23,339,314.07
0	21	469664	1688	0	0.21	0.3	0.49	$34,302,324.88
430	11	0	2032	0.3	0.11	0	0.59	$1,603,718.23
143	11	313109	2032	0.1	0.11	0.2	0.59	$23,529,739.85
0	11	469664	2032	0	0.11	0.3	0.59	$34,492,750.66
430	1	0	2376	0.3	0.01	0	0.69	$1,794,144.00
287	1	156555	2376	0.2	0.01	0.1	0.69	$12,757,154.82
143	1	313109	2376	0.1	0.01	0.2	0.69	$23,720,165.63
0	1	469664	2376	0	0.01	0.3	0.69	$34,683,176.44

let us change the design target to 0.7. In other words, the system design allows withstanding only 70% of the uncertain range between the given upper and lower bounds. Figure 9.19 shows the corresponding results. In the minimum TAC_{system} on the target surface ($630,675.83), FC, PV, and BT provide 60%, 30%, and 10% of total power, respectively, that is, 51 FC, 573 PV, and 310 battery modules.

9.4.4 EFFECTS OF CHANGING THE SUPPLY-TO-DEMAND RATIO

The minimum values of TAC_{system} under the constraint of $FI_d = 1$ or $FI_t = 1$ have been determined according to a series of different r_{SD} between 1.00 and 2.75 (see Tables 9.5 and 9.6). The lowest TAC_{system} in the case of $FI_d = 1$ is $699,943, which

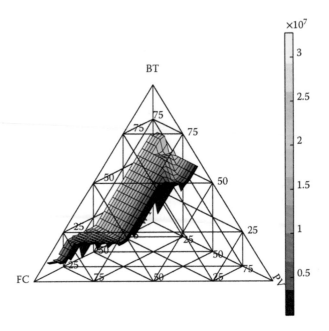

FIGURE 9.18 Surface area that represents all the possible ratios of PV, FC, WT, and battery modules with FI_t value of 1 for $r_{SD} = 2$.

can be obtained at $r_{SD} = 1.75$, whereas this value in the case of $FI_t = 1$ is \$692,858 at also $r_{SD} = 1.75$.

From Figure 9.20, it can be observed that a higher r_{SD} results in an increase in TAC_{system}, which is clearly due to overdesigns in the power-generating units. On the other hand, a lower r_{SD} also leads to a larger total annual cost because more expensive but more stable power sources (such as the fuel cell and battery) are called for in this situation. Notice that the minimum TAC_{system} corresponding to $FI_t = 1$ is lower than that associated with $FI_d = 1$. This result is primarily due to the fact that the battery is more expensive than the fuel cell and, in the present example, the chosen range for the accumulated deviation in power demand makes it possible to find the FI_t surface with smaller battery power ratios. Finally, notice that no feasible designs can be found to achieve $FI_d = 1$ if $r_{SD} < 1.3$ and $FI_t = 1$ if $r_{SD} < 1.25$.

9.4.5 MERITS OF INCORPORATING THE ELECTROLYZER

As indicated before, to design a flexible hybrid power system, it is necessary to ensure feasibility even when its instantaneous power output is at the lowest level. This approach inevitably results in a waste of surplus energy when the total supply from natural resources peaks unexpectedly. In this situation, an electrolyzer can be introduced to generate hydrogen when the battery has no extra capacity for storing electricity. This modified system layout can be found in Figure 9.21. Because the energy efficiency of transforming electricity into hydrogen with the electrolyzer and

TABLE 9.4

Number of Units and Power Ratios Required for $Fl_d = 1$ Where $r_{SD} = 2$

N_{PV}	N_{FC}	N_{WT}	N_B	r_{PV}	r_{FC}	r_{WT}	r_B	TAC ($)
143	80	0	310	0.1	0.81	0.0	0.09	**$693,418.34**
0	80	156555	310	0.0	0.81	0.1	0.09	$11,656,429.15
143	71	0	654	0.1	0.71	0.0	0.19	$883,844.12
0	71	156555	654	0.0	0.71	0.1	0.19	$11,846,854.93
287	61	0	654	0.2	0.61	0.0	0.19	$862,929.62
143	61	156555	654	0.1	0.61	0.1	0.19	$11,825,940.43
0	61	313109	654	0.0	0.61	0.2	0.19	$22,788,951.24
430	51	0	654	0.3	0.51	0.0	0.19	$842,015.11
287	51	156555	654	0.2	0.51	0.1	0.19	$11,805,025.93
143	51	313109	654	0.1	0.51	0.2	0.19	$22,768,036.74
0	51	469664	654	0.0	0.51	0.3	0.19	$33,731,047.55
430	41	0	999	0.3	0.41	0.0	0.29	$1,032,440.89
287	41	156555	999	0.2	0.41	0.1	0.29	$11,995,451.70
143	41	313109	999	0.1	0.41	0.2	0.29	$22,958,462.51
0	41	469664	999	0.0	0.41	0.3	0.29	$33,921,473.33
430	31	0	1343	0.3	0.31	0.0	0.39	$1,222,866.67
287	31	156555	1343	0.2	0.31	0.1	0.39	$12,185,877.48
143	31	313109	1343	0.1	0.31	0.2	0.39	$23,148,888.29
0	31	469664	1343	0.0	0.31	0.3	0.39	$34,111,899.10
430	21	0	1688	0.3	0.21	0.0	0.49	$1,413,292.45
287	21	156555	1688	0.2	0.21	0.1	0.49	$12,376,303.26
143	21	313109	1688	0.1	0.21	0.2	0.49	$23,339,314.07
0	21	469664	1688	0.0	0.21	0.3	0.49	$34,302,324.88
573	11	0	1688	0.4	0.11	0.0	0.49	$1,392,377.94
430	11	156555	1688	0.3	0.11	0.1	0.49	$12,355,388.76
287	11	156555	2032	0.2	0.11	0.1	0.59	$12,566,729.04
143	11	313109	2032	0.1	0.11	0.2	0.59	$23,529,739.85
0	11	469664	2032	0.0	0.11	0.3	0.59	$34,492,750.66
573	1	0	2032	0.4	0.01	0.0	0.59	$1,582,803.72
430	1	156555	2032	0.3	0.01	0.1	0.59	$12,545,814.53
287	1	313109	2032	0.2	0.01	0.2	0.59	$23,508,825.35
143	1	313109	2376	0.1	0.01	0.2	0.69	$23,720,165.63
0	1	469664	2376	0.0	0.01	0.3	0.69	$34,683,176.44

then back to electricity with the fuel cell is only about 40% (Schoenung, 2011), one should generate hydrogen only when needed to avoid conversion loss. Let us assume that a simple cascade control loop (Figure 9.22) is available for manipulating the hydrogen generation rate of the electrolyzer and the hydrogen feed rate of fuel cells.

The primary controller manipulates the set point of the battery's SOC based on the hydrogen pressure in the storage tank. Moreover, the SOC is controlled by adjusting the hydrogen feed rate of the fuel cell and the hydrogen production rate of the electrolyzer. When the SOC is lower than the set point, the feed rate of the fuel cell

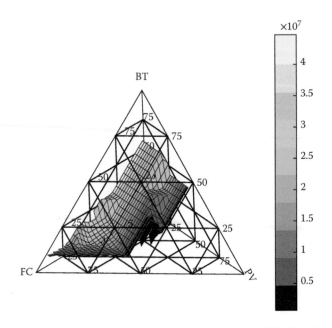

FIGURE 9.19 Surface area that represents all the possible ratios of PV, FC, WT, and battery modules with the FI_d value of 0.7 for $r_{SD} = 2$.

TABLE 9.5

Minimum TAC_{system} and Number of Units for $FI_d = 1$

r_{SD}	N_{PV}	N_{FC}	N_{WT}	N_B	TAC ($)
1.10	N/A	N/A	N/A	N/A	N/A
1.20	N/A	N/A	N/A	N/A	N/A
1.30	0.0	15.3	0.0	1994.0	$1,500,515
1.40	0.0	47.9	0.0	941.0	$930,517
1.50	0.0	68.2	0.0	491.2	$721,633
1.75	37.5	84.5	0.0	331.8	**$699,943**
2.00	143.0	90.9	0.0	310.2	$727,313
2.25	321.7	89.7	0.0	349.0	$760,517
2.50	536.2	85.6	0.0	387.8	$780,897
2.75	786.4	78.7	0.0	426.5	$788,453

is increased to raise the power output, and the hydrogen generation rate of the electrolyzer is decreased to reduce the power consumption. When the hydrogen pressure is higher than its set point, the SOC set point is raised proportionally to increase the hydrogen usage and decrease the hydrogen generation rate.

The capital cost of an electrolyzer is configured at $4,730, and its rated power consumption is 60 kW and the output flow is 10 Nm³/hr (Schoenung, 2011). The capital cost of a high-pressure hydrogen storage tank is set at $2865. It is

TABLE 9.6

Minimum TAC_{system} and Number of Units for $FI_t = 1$

r_{SD}	N_{PV}	N_{FC}	N_{WT}	N_B	TAC ($)
1.10	N/A	N/A	N/A	N/A	N/A
1.20	N/A	N/A	N/A	N/A	N/A
1.25	0.0	17.5	0.0	1615.7	$1,243,844
1.30	0.0	29.2	0.0	1344.3	$1,114,620
1.40	0.0	47.9	0.0	941.0	$930,517
1.50	0.0	68.2	0.0	491.2	$721,633
1.75	12.5	87.4	0.0	301.6	**$692,858**
2.00	143.0	90.9	0.0	310.2	$727,313
2.25	321.7	89.7	0.0	349.0	$760,517
2.50	536.2	85.6	0.0	387.8	$780,897
2.75	786.4	78.7	0.0	426.5	$788,453

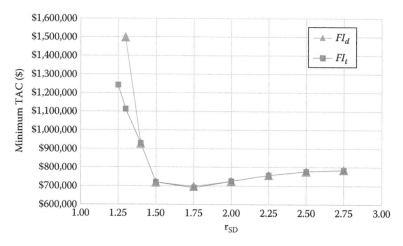

FIGURE 9.20 Minimum total annual costs at different r_{SD}s under the constraint of $FI_d = 1$ or $FI_t = 1$.

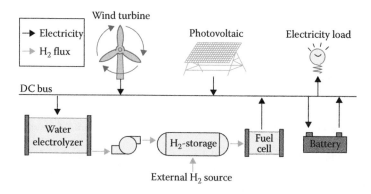

FIGURE 9.21 PV/FC/WT/EL hybrid power generation system layout.

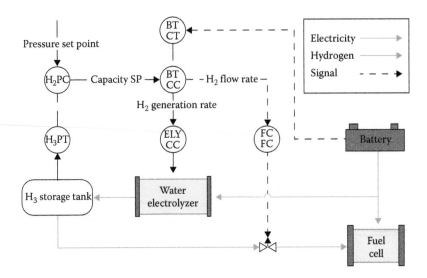

FIGURE 9.22   Cascade control layout.

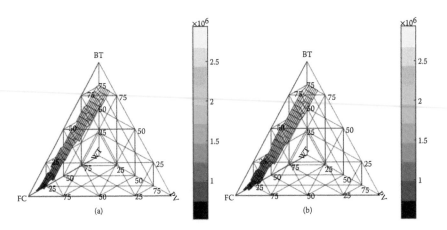

(a)   (b)

FIGURE 9.23   (a) Target surface for $FI_d =$ without electrolyzer. (b) Target surface for $FI_d =$ with electrolyzer.

assumed that the maximum tolerable pressure of this tank is 350 bar and its size is 258 L. Consequently; the total annual cost should be computed as follows:

$$TAC_{system} = TAC_{PV} + TAC_{FC} + TAC_{WT} + TAC_B + TAC_{ELY} + TAC_{tank} \quad (9.49)$$

Let us assign $r_{SD} = 1.5$ and adopt the same weather data as before for flexibility and economic analyses. Figure 9.23 shows the resulting target surface for achieving $FI_d = 1$. Note that the volume of the feasible region does not improve significantly, but TAC_{system} increases about \$6000 due to this additional electrolyzer. Note also that

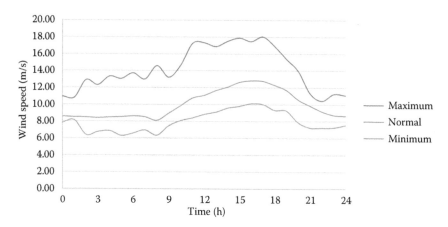

FIGURE 9.24 Hourly sun irradiance profile of Richfield, Idaho, USA.

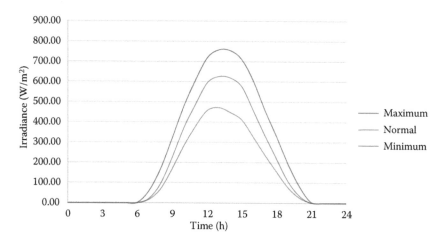

FIGURE 9.25 Hourly wind speed profile of Richfield, Idaho, USA.

the modified hybrid power system is incapable of achieving the originally intended purpose of incorporating an electrolyzer, that is, to store the excess energy provided by wind and/or sun. Because a flexibility analysis focuses upon the worst-case scenario and the lower bounds of wind speed and solar irradiation rate in Annan essentially yield little or even no power output, there is no need for an electrolyzer.

Although the electrolyzer is not needed in Annan, it may still be useful elsewhere. Let us instead consider the weather data at Richfield in Idaho, USA. Figures 9.24 and 9.25 show the time profiles of local solar irradiance and wind speed in 2014, respectively. Let us again set r_{SD} to be 1.5, and the corresponding target surfaces for $FI_d = 1$ are shown in Figure 9.26. The lowest TAC_{system} without the electrolyzer is $699,234, and the corresponding energy ratios are 66% FC (49 units), 3% WT (53 units), 9% PV (98 units), and 22% BT (568 units). This lowest TAC_{system} can be reduced to

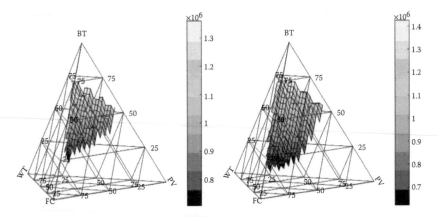

FIGURE 9.26 (a) Target surface for $FI_d = 1$ without electrolyzer. (b) Target surface for $FI_d = 1$ with electrolyzer.

TABLE 9.7

Accumulated Deviations of Three Uncertainties in Richfield, Idaho, USA.

	$\Delta\Theta^+$	$\Delta\Theta^-$	$\int_0^H \Delta\theta^+(\tau)d\tau$	$\int_0^H \Delta\theta^-(\tau)d\tau$
Wind speed (m hr/s)	71.9	35.5	102.7	50.7
Solar irradiance (W hr/m²)	1038.0	889.4	1482.8	1270.6
Power consumption (kWh)	1896.0	3262.4	2404.2	3265.9

$594,116 after adding an electrolyzer. The resulting hybrid system consists of 41 FC units, 53 WT units, 284 PV units, and 413 BT units. Notice that the feasible volume expands toward PV. This is because the nominal solar energy absorbed during the daytime cannot be completely utilized. With the electrolyzer, the surplus energy that exceeds the battery capacity can be transformed into hydrogen. The stored hydrogen can then be utilized later in the fuel cell to generate power when the battery's SOC is low. Also, notice that the feasible volume tends to incorporate a lower portion of BT. This tendency is because of the fact that the electrolyzer acts as a secondary energy storage unit.

Based on the model parameters listed in Table 9.7, the temporal flexibility analysis was also applied to generate the target surface in Figure 9.27. The feasible volume becomes larger due to these smaller accumulated deviations. Furthermore, the lowest TAC_{system} further decreases to $381,398, which is caused by replacing FC units by PV units. Specifically, there are 29 FC units, 36 WT units, 602 PV units, and 103 BT units in this hybrid power system.

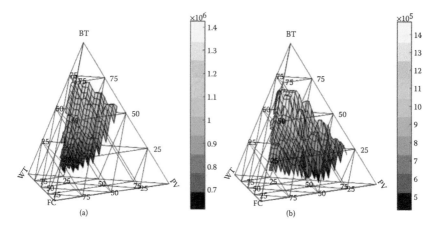

FIGURE 9.27 (a) Target surface for $FI_d = 1$ without electrolyzer. (b) Target surface for $FI_t = 1$ with electrolyzer.

9.5 CONCLUDING REMARKS

In the present chapter, the dynamic and temporal flexibility analyses have been enhanced with the simulation-based optimization strategies. From the results obtained in the benchmark example, it can be observed that the proposed approach saves up to 40% of the computation time. Notice that the maintainability of the corresponding computer code is ensured by its modularity, and this benefit can be clearly demonstrated if more rigorous unit models (e.g., those for the wind turbines and electrolyzers) become available in the future.

Notice also that the dynamic and temporal flexibility analyses have been successfully applied to generate candidate designs of the hybrid power systems in the case studies. A few specific conclusions can be drawn from these studies:

- The dynamic flexibility index should be computed before its temporal counterpart because $FI_d = 1$ is a more stringent requirement, whereas $FI_t = 1$ should be adopted as a design target only under tight budget constraints.
- An optimum energy supply-to-demand ratio (r_{SD}) can be determined by minimizing TAC_{system} with the proposed solution algorithm.
- The electrolyzer is only favorable at locations where there are abundant wind and solar energies.

Although the simulation-based vertex method has been successfully applied in the case studies, there are still a few unresolved research issues. For example, studies are needed to develop more rigorous unit models to facilitate more accurate analyses, more efficient search algorithms to identify system configurations that satisfy the requirements of $FI_d = 1$ and/or $FI_t = 1$, a systematic method to configure proper control schemes, and a logical approach to assess the tradeoff between flexibility and cost.

REFERENCES

Adi, V.S.K., Chang, C.T., 2011. Two-tier search strategy to identify nominal operating conditions for maximum flexibility. *Industrial & Engineering Chemistry Research* 50, 10707–10716.

Adi, V.S.K., Chang, C.T., 2013. A mathematical programming formulation for temporal flexibility analysis. *Computers & Chemical Engineering* 57, 151–158.

Adi, V.S.K., Chang, C.T., 2015. Development of flexible designs for PVFC hybrid power systems. *Renewable Energy* 74, 176–186.

Bajpai, P., Dash, V., 2012. Hybrid renewable energy systems for power generation in stand-alone applications: A review. *Renewable & Sustainable Energy Reviews* 16, 2926–2939.

Banos, R., Manzano-Agugliaro, F., Montoya, F.G., Gil, C., Alcayde, A., Gomez, J., 2011. Optimization methods applied to renewable and sustainable energy: A review. *Renewable & Sustainable Energy Reviews* 15, 1753–1766.

Bansal, V., Perkins, J.D., Pistikopoulos, E.N., 2000. Flexibility analysis and design of linear systems by parametric programming. *Aiche Journal* 46, 335–354.

Bansal, V., Perkins, J.D., Pistikopoulos, E.N., 2002. Flexibility analysis and design using a parametric programming framework. *Aiche Journal* 48, 2851–2868.

Chayawatto, N., Kirtikara, K., Monyakul, V., Jivacate, C., Chenvidhya, D., 2009. DC-AC switching converter modelings of a PV grid-connected system under islanding phenomena. *Renewable Energy* 34, 2536–2544.

Dimitriadis, V.D., Pistikopoulos, E.N., 1995. Flexibility analysis of dynamic-systems. *Industrial & Engineering Chemistry Research* 34, 4451–4462.

Dixon, S.L., Hall, C.A., 2014. *Fluid Mechanics and Thermodynamics of Turbomachinery*, 7th ed. Amsterdam; Boston, MA: Butterworth-Heinemann, animprint of Elsevier, pp. 18, 537.

Erdinc, O., Uzunoglu, M., 2012. Optimum design of hybrid renewable energy systems: Overview of different approaches. *Renewable & Sustainable Energy Reviews* 16, 1412–1425.

Floudas, C.A., Gumus, Z.H., Ierapetritou, M.G., 2001. Global optimization in design under uncertainty: Feasibility test and flexibility index problems. *Industrial & Engineering Chemistry Research* 40, 4267–4282.

Grossmann, I.E., Floudas, C.A., 1987. Active constraint strategy for flexibility analysis in chemical processes. *Computers & Chemical Engineering* 11, 675–693.

Grossmann, I.E., Halemane, K.P., 1982. Decomposition strategy for designing flexible chemical-plants. *Aiche Journal* 28, 686–694.

Halemane, K.P., Grossmann, I.E., 1983. Optimal process design under uncertainty. *Aiche Journal* 29, 425–433.

Hwang, J.J., Chang, W.R., Weng, F.B., Su, A., Chen, C.K., 2008. Development of a small vehicular PEM fuel cell system. *International Journal of Hydrogen Energy* 33, 3801–3807.

Khan, M.J., Iqbal, M.T., 2009. Analysis of a small wind-hydrogen stand-alone hybrid energy system. *Applied Energy* 86, 2429–2442.

Kuo, Y.C., Chang, C.T., 2016. On heuristic computation and application of flexibility indices for unsteady process design. *Industrial & Engineering Chemistry Research* 55, 670–682.

Li, C.H., Zhu, X.J., Cao, G.Y., Sui, S., Hu, M.R., 2009. Dynamic modeling and sizing optimization of stand-alone photovoltaic power systems using hybrid energy storage technology. *Renewable Energy* 34, 815–826.

Lima, F.V., Georgakis, C., 2008. Design of output constraints for model-based non-square controllers using interval operability. *Journal of Process Control* 18, 610–620.

Lima, F.V., Georgakis, C., Smith, J.F., Schnelle, P.D., Vinson, D.R., 2010a. Operability-based determination of feasible control constraints for several high-dimensional nonsquare industrial processes. *Aiche Journal* 56, 1249–1261.

Lima, F.V., Jia, Z., Lerapetritou, M., Georgakis, C., 2010b. Similarities and differences between the concepts of operability and flexibility: The steady-state case. *Aiche Journal* 56, 702–716.

Malcolm, A., Polan, J., Zhang, L., Ogunnaike, B.A., Linninger, A.A., 2007. Integrating systems design and control using dynamic flexibility analysis. *Aiche Journal* 53, 2048–2061.

MathWorks, 2016. *Battery*. The MathWorks, Inc., Massachusetts.

Nelson, D.B., Nehrir, M.H., Wang, C., 2006. Unit sizing and cost analysis of stand-alone hybrid wind/PV/fuel cell power generation systems. *Renewable Energy* 31, 1641–1656.

Ostrovski, G.M., Achenie, L.E.K., Karalapakkam, A.M., Volin, Y.M., 2002. Flexibility analysis of chemical processes: Selected global optimization sub-problems. *Optimization and Engineering* 3, 31–52.

Ostrovsky, G.M., Achenie, L.E.K., Wang, Y.P., Volin, Y.M., 2000. A new algorithm for computing process flexibility. *Industrial & Engineering Chemistry Research* 39, 2368–2377.

Petracci, N.C., Hoch, P.M., Eliceche, A.M., 1996. Flexibility analysis of an ethylene plant. *Computers & Chemical Engineering* 20, S443–S448.

Pistikopoulos, E.N., Grossmann, I.E., 1988a. Evaluation and redesign for improving flexibility in linear-systems with infeasible nominal conditions. *Computers & Chemical Engineering* 12, 841–843.

Pistikopoulos, E.N., Grossmann, I.E., 1988b. Optimal retrofit design for improving process flexibility in linear-systems. *Computers & Chemical Engineering* 12, 719–731.

Pistikopoulos, E.N., Grossmann, I.E., 1989a. Optimal retrofit design for improving process flexibility in nonlinear-systems.1. Fixed degree of flexibility. *Computers & Chemical Engineering* 13, 1003–1016.

Pistikopoulos, E.N., Grossmann, I.E., 1989b. Optimal retrofit design for improving process flexibility in nonlinear-systems.2. Optimal level of flexibility. *Computers & Chemical Engineering* 13, 1087–1096.

Saathoff, S., 2016. Hybrid power is the new green, OSP Magazine.

Sadri, J.A., Hooshmand, P., 2012. A survey study on hybrid photovoltaic system: Technical and economical approach. *Journal of Basic and Applied Scientific Research* 2, 327–333.

Schoenung, S., 2011. *Economic Analysis of Large-Scale Hydrogen Storage for Renewable Utility Applications*. Livermore, CA: Sandia National Laboratories.

Swaney, R.E., Grossmann, I.E., 1985a. An index for operational flexibility in chemical process design.1. Formulation and theory. *Aiche Journal* 31, 621–630.

Swaney, R.E., Grossmann, I.E., 1985b. An index for operational flexibility in chemical process design.2. Computational algorithms. *Aiche Journal* 31, 631–641.

Thomas, C.E., Kuhn, I.F., James, B.D., Lomax, F.D., Baum, G.N., 1998. Affordable hydrogen supply pathways for fuel cell vehicles. *International Journal of Hydrogen Energy* 23, 507–516.

Ulleberg, O., Nakken, T., Ete, A., 2010. The wind/hydrogen demonstration system at Utsira in Norway: Evaluation of system performance using operational data and updated hydrogen energy system modeling tools. *International Journal of Hydrogen Energy* 35, 1841–1852.

Varvarezos, D.K., Grossmann, I.E., Biegler, L.T., 1995. A sensitivity based approach for flexibility analysis and design of linear process systems. *Computers & Chemical Engineering* 19, 1301–1316.

Villalva, M.G., Gazoli, J.R., Ruppert, E., 2009. Comprehensive approach to modeling and simulation of photovoltaic arrays. *IEEE Transactions on Power Electronics* 24, 1198–1208.

Volin, Y.M., Ostrovskii, G.M., 2002. Flexibility analysis of complex technical systems under uncertainty. *Automation and Remote Control* 63, 1123–1136.

Wang, C.S., Nehrir, M.H., 2008. Power management of a stand-alone wind/photovoltaic/fuel cell energy system. *IEEE Transactions on Energy Conversion* 23, 957–967.

Wu, W., Xu, J.P., Hwang, J.J., 2009. Multi-loop nonlinear predictive control scheme for a simplistic hybrid energy system. *International Journal of Hydrogen Energy* 34, 3953–3964.

Wu, W., Zhou, Y.Y., Lin, M.H., Hwang, J.J., 2013. Modeling, design and analysis of a stand-alone hybrid power generation system using solar/urine. *Energy Conversion and Management* 74, 344–352.

Zhou, W., Lou, C.Z., Li, Z.S., Lu, L., Yang, H.X., 2010. Current status of research on optimum sizing of stand-alone hybrid solar-wind power generation systems. *Applied Energy* 87, 380–389.

10 Further Extensions of Flexibility Analyses

This chapter presents two additional areas of technical problems in which flexibility analyses may be beneficial. Because the inherent uncertainties in these problems play critical roles, the flexibility analyses clearly provide useful insights that facilitate the identification of practical solutions.

10.1 LARGE SYSTEM DESIGNS

10.1.1 Polygeneration Processes

Coal and biomass, with their abundant reserves distributed across the world, are promising energy sources for the immediate future. In recent years, several coal-based conversion processes with high utilization efficiency and low carbon dioxide (CO_2) emissions have been developed to replace the current oil-based counterparts, such as the integrated gasification combined cycle (IGCC) and other coal-to-liquids (CTL) processes with carbon capture and sequestration (CCS) (Adams and Barton, 2010; Li and Fan, 2008). However, these single-product technologies cannot easily adapt to the fluctuation of market prices, especially the crude oil price, and their profitability cannot be guaranteed under all economic conditions. Coal- and biomass-based polygeneration processes with multiple products, such as electricity, liquid fuels, and chemicals, have been proposed recently as an alternative. With polygeneration, economic risks can be reduced by diversifying product portfolios, and potentially higher profits can be achieved via optimization. Higher energy efficiency may also be attained in polygeneration processes by tightening heat integration (Liu et al., 2007).

The economic performance of an optimal polygeneration plant for different market prices and carbon taxes has already been studied in a previous work (Chen et al., 2011), and it was demonstrated that such plants could achieve higher net present values (NPVs) than the single-product plants. Although only the designs of static systems with a fixed product mix were investigated, this study indicated that the optimal design of a static system is always close or equal to a single-product system, and hence, the benefit of polygeneration is not significant. In reality, the market prices and demands fluctuate frequently during operation. For example, the prices of liquid fuels (i.e., gasoline and diesel) vary seasonally, and the power prices fluctuate daily. Furthermore, the power prices and demands at peak times can be several times higher than those at off-peak times. Thus, a flexible polygeneration system that can adapt the product mix to match market fluctuations has the potential to achieve good economic performance.

The major challenge in the design of flexible polygeneration systems is to determine the optimal trade-off between flexibility and cost. Although the long-term design problem and the short-term operational problems must both be solved, they are often treated as two separate problems in process design (Chicco and Mancarefla, 2009). For example, Yunt et al. (2008) developed a two-stage optimization formulation for the optimal design of a fuel cell system for varying power demands. Liu et al. (2010a, b) studied the optimal design of a coal polygeneration system coproducing power and methanol with multiple operation periods. In this study, the feedstock and product prices were assumed to increase from period to period due to inflation. The optimal design and operation schedule in three periods (with several years in one period) were determined. However, seasonal variations of market prices and daily fluctuations of power prices, which are critical in flexible polygeneration operations, were not considered. The two-stage formulation was also widely applied to optimal design and operation of many different kinds of systems under uncertainties such as the water networks (Karuppiah and Grossmann, 2008) and the natural gas production networks (Li et al., 2011).

It can be concluded from these discussions that the model formulation of any polygeneration system should incorporate various unexpected disturbances in its upstream and downstream conditions. Although there are a few works on the optimal design and operation of polygeneration systems using flexibility analyses, most of them utilized the aforementioned two-stage formulation. If the design strategy could consider the trade-off between flexibility and capital cost at the same time, the results could be better than those generated by the conventional approach.

10.1.2   Biological Processes

Long before anyone understood the concept of bioreaction, humans were enjoying its benefits. Bread, cheese, wine, and beer were all made possible through what was traditionally known as fermentation. It is the control of such processes that concerns chemical engineers today first and foremost. The scope of bioengineering has grown from simple wine-bottle microbiology to the industrialization of not only food production, but also the production of biotechnology's newer products—antibiotics, enzymes, steroidal hormones, vitamins, sugars, and organic acids. Bioreactors differ from the conventional reactors in that they support and control biological entities. As such, bioreactor systems must be designed to provide a higher degree of control over process upsets and contaminations, because the organisms are more sensitive and less stable than chemicals. Biological organisms, by their nature, will mutate, which may alter the biochemistry of the bioreaction or the physical properties of the organism. Analogous to heterogeneous catalysis, deactivation or mortality occur, and promoters or coenzymes influence the kinetics of the bioreaction. Although the majority of fundamental bioreactor engineering and design issues are similar, maintaining the desired biological activity and eliminating or minimizing undesired activities often present a greater challenge than traditional chemical reactors typically require. As an example, let us consider the industrial-scale A–B–E (acetone–butanol–ethanol) fermentation. Butanol has recently been proposed as a gasoline additive, or even as a complete gasoline replacement (Lee et al., 2008). Note that it is superior to ethanol because it has higher energy content, lower volatility, and less corrosiveness (Lee et al., 2008).

Batch fermentation is the most often used operation mode in industries due to its high efficiency and good control (Cardona and Sanchez, 2007; Ghose and Tyagi, 1979). The fed-batch fermentation is another option and is usually considered when substrate inhibition or catabolite repression might occur (Luli and Strohl, 1990; Modak et al., 1986). Typically, the fed-batch fermentation is started with a low substrate concentration. When the fermentation culture consumes the substrate, more substrate is then added to maintain the fermentation process while not exceeding the detrimental substrate level. Also, the dilution effect during the addition of the substrate solution may solve the problem of catabolite toxicity, such as the acetate problem in *E. coli* fermentation (Luli and Strohl, 1990). The continuous operation is, of course, the third option and has several advantages over the batch and fed-batch alternatives, including minimizing equipment downtime and time loss due to the lag phase of the microbial culture.

Although three different modes are available, the system designs have been mostly developed according to fixed initial conditions. Traditionally, the biochemical products have been manufactured in large facilities with multiple fixed, stainless steel bioreactors ranging in size from 100 L to 20,000 L. Such facilities were often designed for producing a single product or for campaigning just a few. Obviously, designs of these facilities should consider the uncertainties caused by the biological entities. The uncertainties could come in the form of the unknown reaction rates and stoichiometry, mass transfer, heat transfer, and turbulence and mixing on product distribution, as well as operational characteristics. These phenomena need to be expressed in accurate but tractable models that can be used for design and optimization calculations. The equipment sizing should not only rely on simple safety factors, but instead on a rigorous flexibility analysis.

10.1.3 MEMBRANE CASCADES

Downstream processing is an indispensable part of the biotechnological and pharmaceutical production processes. For nearly every product manufactured in these industrial sectors, one starts with a dilute suspension and tries to generate a purified dry product. Most of the downstream processes include four main steps: removal of insoluble particles, isolation of the product, purification, and polishing. The major part of the production costs of pharmaceuticals can be imputed to the downstream processing steps due to their ineffectiveness. In fact, product recovery should be considered the most expensive part of the entire process (Degerman et al., 2008). Several techniques are available for the treatment of the solutes, including distillation, crystallization, chromatography, adsorption–desorption, ion exchange, extraction, molecular imprinting, and membranes (Siew et al., 2013a; Szekely et al., 2013). Most membrane technologies have found room in downstream processing, for example, microfiltration (van Reis and Zydney, 2007), ultrafiltration (Grote et al., 2011), nanofiltration (Rathore and Shirke, 2011), and reverse osmosis (Grote et al., 2012). The solute separation processes by membrane can achieve high product purity and process yield when the solutes to be separated show appreciable differences in their molecular sizes. From the fact that the more challenging separations involving solutes with closer sizes cannot be carried out by the single-stage processes, additional stages become necessary. To this end, it was demonstrated that the membrane cascades can be configured satisfactorily. The design of membrane cascades has been studied extensively in the past (Gunderson et al., 2007; Lightfoot, 2005; Lightfoot et al., 2008). Various types of membranes have been used in a cascade scheme for applications ranging from microfiltration (Abatemarco et al., 1999) to reverse osmosis (Abejon et al., 2012). Ultrafiltration and nanofiltration are the most frequently used membrane technologies for biotechnological and pharmaceutical purposes. The former is widely used for the separation and fractionation of proteins and other biological molecules (Arunkumar and Etzel, 2013; Cheang and Zydney, 2004; Ghosh, 2003; Isa et al., 2007; Mayani et al., 2009, 2010; Mohanty and Ghosh, 2008; Overdevest et al., 2002; Vanneste et al., 2011; Wang et al., 2010), whereas the latter is implemented for pesticide removal (Caus et al., 2009), solvent exchange (Lin and Livingston, 2007), and separation and fractionation of pharmaceutical solutes (Siew et al., 2013a, b; Vanneste et al., 2013).

Although significant progress has been achieved in recent years in designing downstream processing systems, there is still room for improvement by taking advantage of the well-developed process systems engineering tools (Troup and Georgakis, 2013). Some recent experiences show the benefits of using such tools in the pharmaceutical industry (Cervera-Padrell et al., 2012; Gernaey et al., 2012). Efforts have been specifically aimed at the downstream process development (Winkelnkemper and Schembecker, 2010a, b), and the optimization of membrane processes has been investigated vigorously (vanReis and Saksena, 1997; Venkiteshwaran and Belfort, 2010). Despite the large number of studies described earlier, the membrane cascade designs still ignore uncertainties. Most of them were obtained under the assumption that the upstream conditions are fixed. As shown in the flexibility analyses in Chapter 7, a significant range of uncertain parameters could render the membrane modules inoperable. This operational problem could further propagate in a membrane cascade where more than one module is in place.

10.1.4 Simulation-Based Approach to Compute Steady-State and Volumetric Flexibility Indices Using Commercial Software

Because commercial software (e.g., ASPEN PLUS®) is often utilized for the material and energy balance calculations in realistic process designs, the simulation-based approach described in Chapter 9 can be modified to compute the steady-state and volumetric flexibility indices via the proper interface between simulation and optimization platforms (e.g., between ASPEN and GAMS). The conceptual algorithms for FI_s and FI_v are summarized in Figures 10.1 and 10.2, respectively.

Notice that the computation flowchart in Figure 10.1 is only a slightly modified version of Figure 9.6. Specifically, the simulation runs required in the third block of the latter flowchart are now facilitated with ASPEN PLUS® instead. On the other

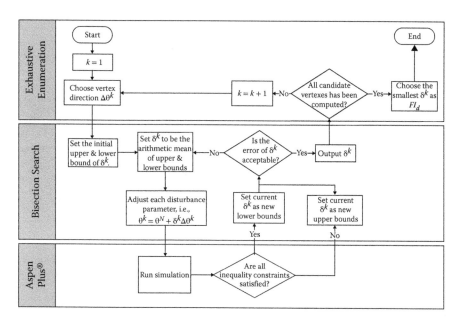

FIGURE 10.1   Simulation-based algorithm for computing steady-state flexibility index.

hand, notice that the computation flowchart of FI_v in Figure 10.2 can be developed on the basis of the algorithms described in Chapter 3. In particular, the following items in Figure 3.5 have been changed:

- Feasibility check performed according to Equation 3.3: ASPEN PLUS® is equipped with an optimization solver where the control variables $\bar{z}$ can be calculated simultaneously. Let us consider the rigorous distillation as an example. For each input, ASPEN PLUS® finds the corresponding control variables (e.g., reboiler load, reflux ratio, etc.) within the specified limits for a distillation column to satisfy the required constraints (e.g., product purity, product flow rate, etc.). If the optimization solver is unable to determine the suitable values for control variables, the corresponding block status should show an error and the resulting value is 1. Otherwise, a status value of 0 indicates that the given condition passes the feasibility check.
- Random line search performed according to Equation 3.4: The bisection search is adopted in this algorithm to locate the proximity points, and because there are two directions along a straight line, it is done twice. Two initial points in each direction are determined. One is placed at the lower/upper limit of uncertain range and the other at **b**. Clearly, the block status associated with each of these points can be evaluated with ASPEN PLUS® also. The third point should be introduced at the middle of the two initial points, and the corresponding block status can also be evaluated in the same way.

On the basis of the block status values generated with ASPEN PLUS®, the bisection search can be carried out to identify the closest two points, where one is marked with a block status of 1 and the other 0. The latter point is then reported as the boundary point.

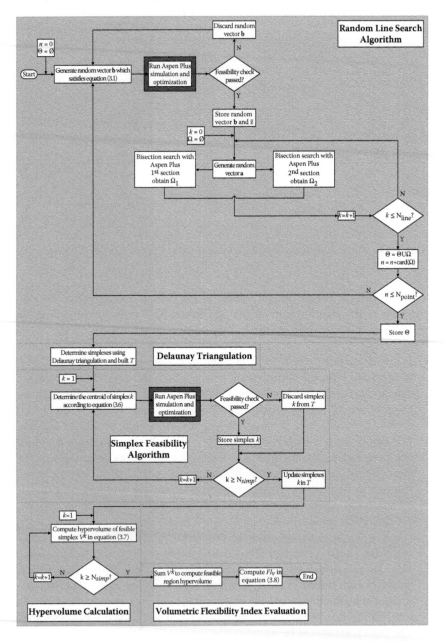

FIGURE 10.2   Simulation-based algorithm for computing volumetric flexibility index.

10.2   PID CONTROLLER DESIGNS

10.2.1   PID CONTROLLER SETTINGS

Traditionally, the PID controller settings can be estimated with a number of alternative techniques, for example, the direct synthesis (DS) method, the internal model control (IMC) method, and the controller tuning relations. Note that each method can be applied in advance on the basis of a given process model. Because the subsequent online tuning can be time consuming, it is very useful to have good initial estimates of controller settings in order to minimize the required time and effort.

In the DS method, the controller settings are determined according to a process model and a desired closed-loop transfer function. The latter is usually specified either for the set-point changes or external disturbances (Chen and Seborg, 2002). This approach yields valuable insight into the connections between the process model and the corresponding controller. Although the resulting feedback controllers do not always have a PID structure, the DS method does produce PI or PID controllers for many simple systems. A more comprehensive model-based method, *Internal Model Control* (IMC), was developed by Morari and coworkers (Garcia and Morari, 1982; Rivera et al., 1986). The IMC method also relies on an assumed model and leads to analytical expressions for the controller settings.

Analytical expressions for PID controller settings have been derived from other perspectives as well. These expressions are referred to as the controller tuning relations, or just tuning relations, and they can be roughly divided into three groups as follows:

1. *IMC tuning relations.* The IMC method can be used to derive PID controller settings for a variety of transfer function models. Different tuning relations can be derived depending on the type of low-pass filter and time-delay approximation that are selected (Chien and Fruehauf, 1990; Rivera et al., 1986; Skogestad, 2003). Accordingly, the PID controller tuning relations for the parallel form could be derived (Chien and Fruehauf, 1990) for common types of process models. Note that the values in the tuning relations table (Chien and Fruehauf, 1990) are all derived based on fixed process conditions, and the value of the tuning parameters are assumed to be constant for the whole operation campaign.
2. *Tuning relations based on integral error criteria.* Controller tuning relations have been developed that optimize the closed-loop response for a simple process model and a specified disturbance or set-point change. The optimum settings minimize an integral error criterion, that is, the integral of the absolute value of the error (IAE), the integral of the squared error (ISE), and the integral of the time-weighted absolute error (ITAE). In general, the ITAE is the more preferred criterion because it usually results in the most conservative controller settings.
3. *Miscellaneous tuning relations.* Two early controller tuning relations were published by Ziegler and Nichols (1993) and Cohen and Coon (1953). These well-known tuning relations were developed to provide closed-loop

responses that have a quarter decay ratio. Because a response with a quarter decay ratio is considered to be excessively oscillatory for most process control applications, these tuning relations are not recommended.

10.2.2 CONTROLLER PERFORMANCE CRITERIA

The intended function of installing a PID controller is to ensure that the closed-loop system has the desired system dynamics and steady-state response characteristics. Ideally, the closed-loop system should meet the following performance criteria:

1. The closed-loop system must be stable.
2. The effects of disturbances are minimized, providing good disturbance rejection.
3. Rapid, smooth responses to set-point changes are obtained, that is, good set-point tracking.
4. Steady-state error (offset) is eliminated.
5. Excessive control action is avoided.
6. The control system is robust, that is, insensitive to changes in process conditions and to inaccuracies in the process model.

In typical control applications, it is not possible to achieve all goals simultaneously because they involve inherent conflicts and tradeoffs. First of all, the selected PID controller must balance two important objectives: performance and robustness. A feedback control system exhibits a high degree of performance if it provides rapid and smooth responses to disturbances and set-point changes with little, if any, oscillation. On the other hand, a control system should also be robust, that is, the controller provides satisfactory performance for a wide range of process conditions and for a reasonable degree of model inaccuracy. Robustness can usually be achieved by choosing conservative controller settings, but this choice tends to result in poor performance.

A second type of tradeoff occurs because PID controller settings that provide excellent disturbance rejection can produce large overshoots for set-point changes. On the other hand, if the controller settings are specified to provide excellent set-point tracking, the disturbance responses can be very sluggish. Thus, a tradeoff between set-point tracking and disturbance rejection usually also occurs for the PID controllers.

10.2.3 POTENTIAL APPLICATION STRATEGY

In a traditional application, the controller settings are set in advance according to a process model that can be expressed in the form of a set of differential equations or transfer functions. However, this controller design approach totally ignores the inherent system characteristics that are expressible in the form of inequalities and, also, the inherent uncertainties in external disturbances. Clearly, the dynamic and/or temporal flexibility analyses may be useful for providing insights to address these important issues. Note also that in the previous chapters, perfect control is assumed in the flexibility analyses. If an explicit control policy is incorporated in the flexibility index model, the corresponding controller settings could be systematically selected according to two additional performance criteria: FI_d and FI_t. More specifically, let us express the general control policy as

$$\mathbf{z} = \mathbf{C}(\mathbf{x}, \boldsymbol{\omega}_c) \qquad (10.1)$$

where $\mathbf{C}$ and $\boldsymbol{\omega}_c$ denote the vectors of controller models and controller settings, respectively. Note that this equation can be directly substituted into Equations 4.1, 4.2, 5.2, and 5.3, and the computation procedures described in Chapters 4 and 5 can still be carried out without any modification.

REFERENCES

Abatemarco, T., Stickel, J., Belfort, J., Frank, B.P., Ajayan, P.M., Belfort, G., 1999. Fractionation of multiwalled carbon nanotubes by cascade membrane microfiltration. *Journal of Physical Chemistry B* 103, 3534–3538.

Abejon, R., Garea, A., Irabien, A., 2012. Optimum design of reverse osmosis systems for hydrogen peroxide ultrapurification. *Aiche Journal* 58, 3718–3730.

Adams, T.A., Barton, P.I., 2010. High-efficiency power production from coal with carbon capture. *Aiche Journal* 56, 3120–3136.

Arunkumar, A., Etzel, M.R., 2013. Fractionation of alpha-lactalbumin from beta-lactoglobulin using positively charged tangential flow ultrafiltration membranes. *Separation and Purification Technology* 105, 121–128.

Cardona, C.A., Sanchez, O.J., 2007. Fuel ethanol production: Process design trends and integration opportunities. *Bioresource Technology* 98, 2415–2457.

Caus, A., Vanderhaegen, S., Braeken, L., Van der Bruggen, B., 2009. Integrated nanofiltration cascades with low salt rejection for complete removal of pesticides in drinking water production. *Desalination* 241, 111–117.

Cervera-Padrell, A.E., Skovby, T., Kiil, S., Gani, R., Gernaey, K.V., 2012. Active pharmaceutical ingredient (API) production involving continuous processes—A process systems engineering (PSE)-assisted design framework. *European Journal of Pharmaceutics and Biopharmaceutics* 82, 437–456.

Cheang, B.L., Zydney, A.L., 2004. A two-stage ultrafiltration process for fractionation of whey protein isolate. *Journal of Membrane Science* 231, 159–167.

Chen, D., Seborg, D.E., 2002. PI/PID controller design based on direct synthesis and disturbance rejection. *Industrial & Engineering Chemistry Research* 41, 4807–4822.

Chen, Y., Adams, T.A., Barton, P.I., 2011. Optimal design and operation of static energy polygeneration systems. *Industrial & Engineering Chemistry Research* 50, 5099–5113.

Chicco, G., Mancarefla, P., 2009. Distributed multi-generation: A comprehensive view. *Renewable & Sustainable Energy Reviews* 13, 535–551.

Chien, I.L., Fruehauf, P.S., 1990. Consider IMC tuning to improve controller performance. *Chemical Engineering Progress* 86, 33–41.

Cohen, G.H., Coon, G.A., 1953. Theoretical considerations of retarded control. *Journal of Dynamic Systems Measurement and Control-Transactions of the Asme* 75, 827–834.

Degerman, M., Jakobsson, N., Nilsson, B., 2008. Designing robust preparative purification processes with high performance. *Chemical Engineering & Technology* 31, 875–882.

Garcia, C.E., Morari, M., 1982. Internal model control. 1. A unifying review and some new results. *Industrial & Engineering Chemistry Process Design and Development* 21, 308–323.

Gernaey, K.V., Cervera-Padrell, A.E., Woodley, J.M., 2012. A perspective on PSE in pharmaceutical process development and innovation. *Computers & Chemical Engineering* 42, 15–29.

Ghose, T.K., Tyagi, R.D., 1979. Rapid ethanol fermentation of cellulose hydrolysate. 1. Batch versus continuous systems. *Biotechnology and Bioengineering* 21, 1387–1400.

Ghosh, R., 2003. Novel cascade ultrafiltration configuration for continuous, high-resolution protein-protein fractionation: A simulation study. *Journal of Membrane Science* 226, 85–99.

Grote, F., Frohlich, H., Strube, J., 2011. Integration of ultrafiltration unit operations in biotechnology process design. *Chemical Engineering & Technology* 34, 673–687.

Grote, F., Frohlich, H., Strube, J., 2012. Integration of reverse-osmosis unit operations in biotechnology process design. *Chemical Engineering & Technology* 35, 191–197.

Gunderson, S.S., Brower, W.S., O'Dell, J.L., Lightfoot, E.N., 2007. Design of membrane cascades. *Separation Science and Technology* 42, 2121–2142.

Isa, M.H.M., Coraglia, D.E., Frazier, R.A., Jauregi, P., 2007. Recovery and purification of surfactin from fermentation broth by a two-step ultrafiltration process. *Journal of Membrane Science* 296, 51–57.

Karuppiah, R., Grossmann, I.E., 2008. Global optimization of multiscenario mixed integer nonlinear programming models arising in the synthesis of integrated water networks under uncertainty. *Computers & Chemical Engineering* 32, 145–160.

Lee, S.Y., Park, J.H., Jang, S.H., Nielsen, L.K., Kim, J., Jung, K.S., 2008. Fermentative butanol production by clostridia. *Biotechnology and Bioengineering* 101, 209–228.

Li, F.X., Fan, L.S., 2008. Clean coal conversion processes - progress and challenges. *Energy & Environmental Science* 1, 248–267.

Li, X., Armagan, E., Tomasgard, A., Barton, P.I., 2011. Stochastic pooling problem for natural gas production network design and operation under uncertainty. *Aiche Journal* 57, 2120–2135.

Lightfoot, E.N., 2005. Can membrane cascades replace chromatography? Adapting binary ideal cascade theory of systems of two solutes in a single solvent. *Separation Science and Technology* 40, 739–756.

Lightfoot, E.N., Root, T.W., O'Dell, J.L., 2008. Emergence of ideal membrane cascades for downstream processing. *Biotechnology Progress* 24, 599–605.

Lin, J.C.T., Livingston, A.G., 2007. Nanofiltration membrane cascade for continuous solvent exchange. *Chemical Engineering Science* 62, 2728–2736.

Liu, P., Gerogiorgis, D.I., Pistikopoulos, E.N., 2007. Modeling and optimization of polygeneration energy systems. *Catalysis Today* 127, 347–359.

Liu, P., Pistikopoulos, E.N., Li, Z., 2010a. Decomposition based stochastic programming approach for polygeneration energy systems design under uncertainty. *Industrial & Engineering Chemistry Research* 49, 3295–3305.

Liu, P., Pistikopoulos, E.N., Li, Z., 2010b. A multi-objective optimization approach to polygeneration energy systems design. *Aiche Journal* 56, 1218–1234.

Luli, G.W., Strohl, W.R., 1990. Comparison of growth, acetate production, and acetate inhibition of escherichia-coli strains in batch and fed-batch fermentations. *Applied and Environmental Microbiology* 56, 1004–1011.

Mayani, M., Filipe, C.D.M., Ghosh, R., 2010. Cascade ultrafiltration systems-Integrated processes for purification and concentration of lysozyme. *Journal of Membrane Science* 347, 150–158.

Mayani, M., Mohanty, K., Filipe, C., Ghosh, R., 2009. Continuous fractionation of plasma proteins HSA and HIgG using cascade ultrafiltration systems. *Separation and Purification Technology* 70, 231–241.

Modak, J.M., Lim, H.C., Tayeb, Y.J., 1986. General-characteristics of optimal feed rate profiles for various fed-batch fermentation processes. *Biotechnology and Bioengineering* 28, 1396–1407.

Mohanty, K., Ghosh, R., 2008. Novel tangential-flow countercurrent cascade ultrafiltration configuration for continuous purification of humanized monoclonal antibody. *Journal of Membrane Science* 307, 117–125.

Overdevest, P.E.M., Hoenders, M.H.J., van't Riet, K., van der Padt, A., Keurentjes, J.T.F., 2002. Enantiomer separation in a cascaded micellar-enhanced ultrafiltration system. *Aiche Journal* 48, 1917–1926.

Rathore, A.S., Shirke, A., 2011. Recent developments in membrane-based separations in bio-technology processes: Review. *Preparative Biochemistry & Biotechnology* 41, 398–421.

Rivera, D.E., Morari, M., Skogestad, S., 1986. Internal model control. 4. PID controller-design. *Industrial & Engineering Chemistry Process Design and Development* 25, 252–265.

Siew, W.E., Livingston, A.G., Ates, C., Merschaert, A., 2013a. Continuous solute fractionation with membrane cascades - A high productivity alternative to diafiltration. *Separation and Purification Technology* 102, 1–14.

Siew, W.E., Livingston, A.G., Ates, C., Merschaert, A., 2013b. Molecular separation with an organic solvent nanofiltration cascade - augmenting membrane selectivity with process engineering. *Chemical Engineering Science* 90, 299–310.

Skogestad, S., 2003. Simple analytic rules for model reduction and PID controller tuning. *Journal of Process Control* 13, 291–309.

Szekely, G., Gil, M., Sellergren, B., Heggie, W., Ferreira, F.C., 2013. Environmental and eco-nomic analysis for selection and engineering sustainable API degenotoxification processes. *Green Chemistry* 15, 210–225.

Troup, G.M., Georgakis, C., 2013. Process systems engineering tools in the pharmaceutical industry. *Computers & Chemical Engineering* 51, 157–171.

van Reis, R., Zydney, A., 2007. Bioprocess membrane technology. *Journal of Membrane Science* 297, 16–50.

Vanneste, J., De Ron, S., Vandecruys, S., Soare, S.A., Darvishmanesh, S., Van der Bruggen, B., 2011. Techno-economic evaluation of membrane cascades relative to simulated moving bed chromatography for the purification of mono- and oligosaccharides. *Separation and Purification Technology* 80, 600–609.

Vanneste, J., Ormerod, D., Theys, G., Van Gool, D., Van Camp, B., Darvishmanesh, S., Van der Bruggen, B., 2013. Towards high resolution membrane-based pharmaceutical sepa-rations. *Journal of Chemical Technology and Biotechnology* 88, 98–108.

vanReis, R., Saksena, S., 1997. Optimization diagram for membrane separations. *Journal of Membrane Science* 129, 19–29.

Venkiteshwaran, A., Belfort, G., 2010. Process optimization diagrams for membrane microfil-tration. *Journal of Membrane Science* 357, 105–108.

Wang, L., Khan, T., Mohanty, K., Ghosh, R., 2010. Cascade ultrafiltration bioreactor-separa-tor system for continuous production of F(ab')(2) fragment from immunoglobulin G. *Journal of Membrane Science* 351, 96–103.

Winkelnkemper, T., Schembecker, G., 2010a. Purification fingerprints for experimentally based systematic downstream process development. *Separation and Purification Technology* 71, 356–366.

Winkelnkemper, T., Schembecker, G., 2010b. Purification performance index and separation cost indicator for experimentally based systematic downstream process development. *Separation and Purification Technology* 72, 34–39.

Yunt, M., Chachuat, B., Mitsos, A., Barton, P.I., 2008. Designing man-portable power genera-tion systems for varying power demand. *Aiche Journal* 54, 1254–1269.

Ziegler, J.G., Nichols, N.B., 1993. Optimum settings for automatic controllers. *Journal of Dynamic Systems Measurement and Control-Transactions of the Asme* 115, 220–222.

Index

Printed and bound by CPI Group (UK) Ltd, Croydon, CR0 4YY

01/11/2024

01782619-0003